Ioannis Hatzilygeroudis and Jim Prentzas (Eds.)

Combinations of Intelligent Methods and Applications

Smart Innovation, Systems and Technologies 8

Editors-in-Chief

Prof. Robert J. Howlett
KES International
PO Box 2115
Shoreham-by-sea
BN43 9AF
UK
E-mail: rjhowlett@kesinternational.org

Prof. Lakhmi C. Jain
School of Electrical and Information Engineering
University of South Australia
Adelaide, Mawson Lakes Campus
South Australia SA 5095
Australia
E-mail: Lakhmi.jain@unisa.edu.au

Further volumes of this series can be found on our homepage: springer.com

Vol. 1. Toyoaki Nishida, Lakhmi C. Jain, and Colette Faucher (Eds.)
Modeling Machine Emotions for Realizing Intelligence, 2010
ISBN 978-3-642-12603-1

Vol. 2. George A. Tsihrintzis, Maria Virvou, and Lakhmi C. Jain (Eds.)
Multimedia Services in Intelligent Environments – Software Development Challenges and Solutions, 2010
ISBN 978-3-642-13354-1

Vol. 3. George A. Tsihrintzis and Lakhmi C. Jain (Eds.)
Multimedia Services in Intelligent Environments – Integrated Systems, 2010
ISBN 978-3-642-13395-4

Vol. 4. Gloria Phillips-Wren, Lakhmi C. Jain, Kazumi Nakamatsu, and Robert J. Howlett (Eds.)
Advances in Intelligent Decision Technologies – Proceedings of the Second KES International Symposium IDT 2010, 2010
ISBN 978-3-642-14615-2

Vol. 5. Robert J. Howlett (Ed.)
Innovation through Knowledge Transfer, 2010
ISBN 978-3-642-14593-3

Vol. 6. George A. Tsihrintzis, Ernesto Damiani, Maria Virvou, Robert J. Howlett, and Lakhmi C. Jain (Eds.)
Intelligent Interactive Multimedia Systems and Services, 2010
ISBN 978-3-642-14618-3

Vol. 7. Robert J. Howlett, Lakhmi C. Jain, and Shaun H. Lee (Eds.)
Sustainability in Energy and Buildings, 2010
ISBN 978-3-642-17386-8

Vol. 8. Ioannis Hatzilygeroudis and Jim Prentzas (Eds.)
Combinations of Intelligent Methods and Applications, 2010
ISBN 978-3-642-19617-1

Ioannis Hatzilygeroudis and Jim Prentzas (Eds.)

Combinations of Intelligent Methods and Applications

Proceedings of the 2nd International Workshop, CIMA 2010, France, October 2010

Ioannis Hatzilygeroudis
Graphics, Multimedia & GIS Lab
Department of Computer Engineering & Informatics
University of Patras
26500 Patras, Hellas, Greece
E-mail: ihatz@ceid.upatras.gr

Jim Prentzas
Democritus University of Thrace
School of Education Sciences
Dept. of Education Sciences in Pre-School Age, Nea Chili
68100 Alexandroupolis, Greece
E-mail: dprentza@psed.duth.gr

ISBN 978-3-642-43597-3 ISBN 978-3-642-19618-8 (eBook)

DOI 10.1007/978-3-642-19618-8

Smart Innovation, Systems and Technologies ISSN 2190-3018

Softcover reprint of the hardcover 1st edition 2011

Typesetting: Scientific Publishing Services Pvt. Ltd., Chennai, India.

Printed on acid-free paper

9 8 7 6 5 4 3 2 1

springer.com

Preface

The combination of different intelligent methods is a very active research area in Artificial Intelligence (AI). The aim is to create integrated or hybrid methods that benefit from each of their components. It is generally believed that complex problems can be easier solved with such integrated or hybrid methods.

Some of the existing efforts combine what are called soft computing methods (fuzzy logic, neural networks and genetic algorithms) either among themselves or with more traditional AI methods such as logic and rules. Another stream of efforts integrates case-based reasoning or machine learning with soft-computing or traditional AI methods. Yet another integrates agent-based approaches with logic and also non-symbolic approaches. Some of the combinations have been quite important and more extensively used, like neuro-symbolic methods, neuro-fuzzy methods and methods combining rule-based and case-based reasoning. However, there are other combinations that are still under investigation, such as those related to the Semantic Web. In some cases, combinations are based on first principles, whereas in other cases they are created in the context of specific applications.

The 2nd Workshop on "Combinations of Intelligent Methods and Applications" (CIMA 2010) was intended to become a forum for exchanging experience and ideas among researchers and practitioners who are dealing with combining intelligent methods either based on first principles or in the context of specific applications.

Important issues of the Workshop were (but not limited to) the following:

- Case-Based Reasoning Integrations
- Genetic Algorithms Integrations
- Combinations for the Semantic Web
- Combinations and Web Intelligence
- Combinations and Web Mining
- Fuzzy-Evolutionary Systems
- Hybrid deterministic and stochastic optimisation methods
- Hybrid Knowledge Representation Approaches/Systems
- Hybrid and Distributed Ontologies
- Information Fusion Techniques for Hybrid Intelligent Systems
- Integrations of Neural Networks
- Intelligent Agents Integrations

- Machine Learning Combinations
- Neuro-Fuzzy Approaches/Systems
- Applications of Combinations of Intelligent Methods to
  - Biology & Bioinformatics
  - Education & Distance Learning
  - Medicine & Health Care

CIMA 2010 was held in conjunction with the 22nd IEEE International Conference on Tools with Artificial Intelligence (ICTAI 2010). Also, we organized a special track in ICTAI 2010, under the same title.

This volume includes revised versions of the papers presented in CIMA 2010 and one of the short papers presented in the corresponding ICTAI 2010 special track. We have also included a paper of ours as invited paper.

We would like to express our appreciation to all authors of submitted papers as well as to the members of CIMA-10 program committee for their excellent work. We would like also to thank Prof. Eric Gregoire, the ICTAI-10 PC Chair for his help and hospitality.

We hope that these proceedings will be useful to both researchers and developers. Given the success of the first two Workshops on combinations of intelligent methods, we intend to continue our effort in the coming years.

Ioannis Hatzilygeroudis
Jim Prentzas

Workshop Organization

Chairs-Organizers

Ioannis Hatzilygeroudis	University of Patras, Greece
Jim Prentzas	Democritus University of Thrace, Greece

Program Committee

Ajith Abraham	MIR Labs, Europe
Plamen Agelov	Lancaster University, UK
Emilio Corchado	University of Salamanca, Spain
Ronald Denaux	University of Leeds, UK
George Dounias	University of the Aegean, Greece
Artur S. d'Avila Garcez	City University, UK
Elpida Keravnou-Papailiou	University of Cyprus, Cyprus
Constantinos Koutsojannis	University of Patras, Greece
Rudolf Kruse	University of Magdeburg, Germany
George Magoulas	Birkbeck College, Univ. of London, UK
Toni Moreno	University Rovira i Virgili, Spain
Ciprian-Daniel Neagu	University of Bradford, UK
Vasile Palade	Oxford University, UK
David Sanchez	University Rovira i Virgili, Spain
Douglas Vieira	Enacom-Handcrafted Technologies, Brazil

Contents

Defeasible Planning through Multi-agent Argumentation

Sergio Pajares and Eva Onaindia

Abstract. The work reported here introduces DefPlanner, an argumentation-based partial-order planner where different agents that have a partial, and possibly contradictory, knowledge of the world articulate arguments for and against supporting preconditions of the actions to be included in a plan. In this paper, we introduce an extension to multiple agents of the defeasible argumentation formalism that has been proposed to address the task of planning in a single agent environment.

1 Introduction

Planning is the art of building control algorithms that synthesize a course of action to achieve a desired set of goals. The mainstream in planning is that of using heuristic functions to evaluate goals and choices of action or states on the basis of their expected utility to the planning agent [7]. In classical planning, intelligent agents must be able to set goals and achieve them, they have a perfect and complete knowledge of the world, and they assume their view of the world can only be changed through the execution of the planning actions. However, in many real-world applications, agents often have contradictory information about the environment and their deductions are not always certain information, but *plausible*, since the conclusions can be withdrawn when new pieces of knowledge are posted by other agents.

On the other hand, argumentation, which has recently become a very active research field in computer science [2], can be viewed as a powerful tool for reasoning about inconsistent information through a rational interaction of arguments for and against some conclusion. Systems that build on defeasible argumentation apply theoretical reasoning for the generation and evaluation of arguments, and they

Sergio Pajares · Eva Onaindia
Universidad Politécnica de Valencia, Camino de Vera s/n 46022 Valencia, Spain
e-mail: spajares@dsic.upv.es, onaindia@dsic.upv.es

I. Hatzilygeroudis and J. Prentzas (Eds.): Comb. of Intell. Methods and Appl., SIST 8, pp. 1–19.
springerlink.com

are used to build applications that deal with incomplete and contradictory information in dynamic domains ([11][5][10][12]). Particularly, the application of an argumentation-based formalism to deal with the defeasible nature of reasoning during the construction of a plan has been addressed by Garcia and Simari [13][6].

This paper extends the work of [6] and presents DefPlanner, a defeasible argumentation planner developed for multi-agent environments. We explicitly consider several entities (agents) in the argumentative process for the support of the conditions of a planning action. Some recent works like [16][15] realize argumentation in multi-agent systems using defeasible reasoning but they are not particularly concerned with the task of planning. Specifically, we consider propositional STRIPS planning representation augmented with the incorporation of different sources of defeasible information (agents). Defplanner is a partial-order planner ([1][9]) that invokes an argumentation process where many different agents with different opinions exchange arguments and counterarguments in order to determine whether a given precondition of an action is supported or not, i.e. it can be defeasibly derived or not.

This paper is organized as follows. Next section summarizes the main notions on defeasible logic and partial-order planning. Section 3 elaborates on the use of argumentation during the construction of a partial-order plan. Section 4 presents the defeasible argumentation process in a multi-agent system, and section 5 presents an example of application. Finally, section 6 concludes and presents some future work.

2 Background

2.1 Defeasible Logic

In this section, we summarize the main concepts of the work on Defeasible Logic Programming (DeLP), a formalism that combines Logic Programming and Defeasible Argumentation [5]. The basic elements in DeLP are facts and rules. Let $\mathscr{L}$ denote a set of literals, where a literal h is a fact A or a negated fact $\sim A$, and, the symbol $\sim$ represents the strong negation. The set of rules is divided into *strict rules*, i.e. rules encoding strict consequences, and *defeasible rules*, which derive uncertain or defeasible conclusions. A strict rule is an ordered pair $head \leftarrow body$, and a defeasible rule is an ordered pair $head \prec body$, where *head* is a literal, and *body* is a finite non-empty set of literals. For example, the strict rule $animal \leftarrow bird$ is denoting the piece of information "a bird is an animal". However, a defeasible rule is used to describe tentative knowledge that may be used if nothing else can be posed against it, e.g. "birds fly" (fly $\prec$ bird).

Using facts, strict and defeasible rules, an agent is able to satisfy some literal h as in other rule-based systems. Let X be a set of facts in $\mathscr{L}$, STR a set of strict rules, and DEF a set of defeasible rules. A **defeasible derivation** for a literal h from X, denoted as $X \mid\!\sim h$, consists of a finite sequence $h_1, \ldots, h_n = h$ of literals such that h_i is a fact ($h_i \in \mathscr{L}$), or there is a rule in $STR \cup DEF$ with head h_i and body $b_1, \ldots, b_k$, and every literal of the body is an element h_j of the sequence appearing before h_i

($j < i$). A set X is contradictory, denoted $X \mid\!\sim \bot$, if two contradictory literals, eg. h and $\sim h$, can be derived from X.

In our planning framework, the agent's knowledge base is formed by a consistent set of facts Ψ, and a set of defeasible rules Δ.

Definition 1. *Let h be a literal, and let $\mathscr{K} = (\Psi, \Delta)$ be the knowledge base of an agent. We say that $\langle \mathscr{A}, h \rangle$ is an* ***argument structure*** *for h, or simply* ***argument*** *for h, if $\mathscr{A}$ is a set of defeasible rules of Δ, such that:*

- *there exists a defeasible derivation of h from $\Psi \cup \mathscr{A}$,*
- *the set $\Psi \cup \mathscr{A}$ is non-contradictory, and*
- *$\mathscr{A}$ is minimal, i.e., there is not a $\mathscr{A}' \subset \mathscr{A}$, such that $\mathscr{A}'$ satisfies the above two conditions.*

The literal h is called the conclusion of the argument, and $\mathscr{A}$ the support of the argument.

Definition 2. *Two literals h_1 and h_2* ***disagree*** *iff the set $\Psi \cup \{h_1, h_2\}$ is contradictory. Two complementary literals h and $\sim h$ disagree because for any set Ψ, $\Psi \cup \{h, \sim h\}$ is contradictory. We say that the argument $\langle \mathscr{A}_1, h_1 \rangle$ is in conflict or counter-argues the argument $\langle \mathscr{A}_2, h_2 \rangle$ at the literal h, if and only if there exists a sub-argument $\langle \mathscr{A}, h \rangle$ of $\langle \mathscr{A}_2, h_2 \rangle$, that is $\mathscr{A} \subseteq \mathscr{A}_2$, such that h and h_1 disagree. If $\langle \mathscr{A}_1, h_1 \rangle$ is a counterargument for $\langle \mathscr{A}_2, h_2 \rangle$ at literal h, then h is called a counter-argument point, and the subargument $\langle \mathscr{A}, h \rangle$ is called the disagreement subargument [5].*

In short, two arguments are in conflict if they support contradictory conclusions, or one of the arguments is in conflict with an inner part of the other argument. That is, if the head of a defeasible rule in one of the arguments contradicts the head of a defeasible rule in the other argument.

In order to deal with counterarguments, a central aspect is to establish a formal **comparison criterion** among arguments. A possible preference relation among arguments is the so-called *generalized specificity* [14]. We consider an argument $\mathscr{A}1$ is preferred to an argument $\mathscr{A}2$ if $\mathscr{A}1$ is more precise (it is based on more information), or more concise (it uses fewer rules in the conclusion derivation). In such a case, it is said $\mathscr{A}1$ is more specific than $\mathscr{A}2$. For example, $\langle \{c \prec a, b\}, c \rangle$ is more specific than $\langle \{\sim c \prec \sim a\}, \sim c \rangle$. We use $\langle \mathscr{A}1, h1 \rangle \succ \langle \mathscr{A}2, h2 \rangle$ to denote $\langle \mathscr{A}1, h1 \rangle$ is more specific than $\langle \mathscr{A}2, h2 \rangle$ The preference criterion is needed to decide whether an argument defeats another or not, as disagreement does not imply preference.

Definition 3. *The argument $\langle A_1, h_1 \rangle$ is a* ***defeater*** *for $\langle A_2, h_2 \rangle$ iff there is a subargument $\langle A, h \rangle$ of $\langle A_2, h_2 \rangle$ such that $\langle A_1, h_1 \rangle$ is a counterargument of $\langle A_2, h_2 \rangle$ at literal h, and $\langle A_1, h_1 \rangle \succ \langle A, h \rangle$.*

Definition 4. *An* ***argumentation line*** *for $\langle \mathscr{A}_0, h_0 \rangle$ is a sequence of arguments, denoted $\Lambda = [\langle \mathscr{A}_0, h_0 \rangle, \ldots, \langle \mathscr{A}_m, h_m \rangle]$, where each element of the sequence $\langle \mathscr{A}_i, h_i \rangle$, $i > 0$, is a defeater of its predecessor $\langle \mathscr{A}_{i-1}, h_{i-1} \rangle$. Certain constraints over Λ are considered in [5] in order to avoid several problematic and undesirable situations that may arise in Λ.*

Definition 5. *A* ***dialectical tree*** *for the argument* $\langle \mathscr{A}_0, h_0 \rangle$*, denoted* $\mathscr{T}_{\langle \mathscr{A}_0, h_0 \rangle}$*, is defined by the root of the tree, labeled with* $\langle \mathscr{A}_0, h_0 \rangle$*, and a set of argumentation lines from the root, where every node (except the root) represents a defeater of its parent, and leaves correspond to non-defeated arguments, arguments with no defeaters.*

Some examples of dialectical trees can be found in [5]. In order to decide whether the argument at the root of a given dialectical tree is defeated or not, it is necessary to perform a bottom-up analysis of the tree. Every leaf of the tree is marked *undefeated* and every inner node is marked *defeated*, if it has at least one child node marked undefeated. Otherwise, it is marked undefeated. Let $\mathscr{T}^*_{\langle \mathscr{A}, h \rangle}$ denote a marked dialectical tree of the argument $\langle \mathscr{A}, h \rangle$. A literal h is said to be ***warranted***, if and only if there is an argument $\langle \mathscr{A}, h \rangle$ for h such that the root of the marked dialectical tree $\mathscr{T}^*_{\langle \mathscr{A}, h \rangle}$ is marked undefeated. In such a case, $\langle \mathscr{A}, h \rangle$ is a warrant for h. If a literal h is a fact then h is also warranted as there are no counterarguments for $\langle \emptyset, h \rangle$. Otherwise, if all arguments for h are marked as defeated then the literal h is said to be ***not warranted***.

2.2 Partial-Order Planning

Planning is the art of building control algorithms that synthesize a course of action to achieve a desired set of goals. We consider planning problems encoded in a formal, first-order language such as STRIPS [4], particularly in a propositional version of STRIPS. We will denote the set of all propositions by $\mathscr{P}$ (ground facts or literals). A planning state s is defined as a finite set propositions $s \subseteq \mathscr{P}$. A (grounded) planning task is a triple $\mathscr{T} = \langle \mathscr{O}, \mathrm{i}, \mathscr{G} \rangle$, where $\mathscr{O}$ is the set of deterministic actions of the agent's model that describes the state changes, and $\mathrm{i} \subseteq \mathscr{P}$ (the initial state) and $\mathscr{G} \subseteq \mathscr{P}$ (the goals) are sets of propositions. An action $a \in \mathscr{O}$ is a tuple $a = (pre(a), add(a), del(a))$, where $pre(a) \subseteq \mathscr{P}$ is the set of propositions that represents the action's preconditions, and $add(a) \subseteq \mathscr{P}$ and $del(a) \subseteq \mathscr{P}$ are the sets of propositions that represent the positive and negative effects, respectively. We will represent an action a as follows:

$$\{q_1, \ldots, q_n, \sim r_1, \ldots, \sim r_m\} \stackrel{id}{\longleftarrow} \{p_1, \ldots, p_k\} \tag{1}$$

where id is the action name, $\forall_{i=1}^{k} p_i \in pre(a)$, $\forall_{i=1}^{n} q_i \in add(a)$, and $\forall_{i=1}^{m} r_i \in del(a)$. An action a is executable in state s if $pre(a) \subseteq s$. The state resulting from executing a is defined as $s' = (s \setminus del(a)) \cup add(a)$. That is, we delete any proposition in s that belongs to $del(a)$, and add the propositions in $add(a)$. A solution plan (Π) for a planning task $\mathscr{T}$ is a set of actions $\Pi = \{a_1, \ldots, a_n\} \subseteq \mathscr{O}$ such that when applied to i, it leads to a final state in which the goals $\mathscr{G}$ are satisfied. A planning task $\mathscr{T}$ is solvable if there exists at least one plan for it.

In what follows, we provide a brief introduction to the Partial-Order Planning (POP) paradigm ([1][9]). A more detailed tutorial can be found in [17]. In POP, search is done through the space of incomplete partially-ordered plans as opposite to state-based planning. Thus, a key concept in POP is that of *partial-order plan*.

Definition 6. *A **partial-order plan** is a tuple* $\Pi = \langle \mathscr{AP}, \mathscr{OR}, \mathscr{CL}, \mathscr{OC}, \mathscr{UL} \rangle$*, where:*

- $\mathscr{AP} \subseteq \mathscr{O}$ *is the set of ground actions*[1] *in* Π.
- $\mathscr{OR}$ *is a set of ordering constraints* ($\prec$) *over* $\mathscr{O}$
- $\mathscr{CL}$ *is a set of causal links over* $\mathscr{O}$*. A causal link is of the form* (a_i, p, a_j)*, and denotes that the precondition* p *of action* a_j *will be supported by an add effect of action* a_i.
- $\mathscr{OC}$ *is the set of* open conditions *of* Π*. Let* $a_i \in \mathscr{O}$*; if* $\exists p \in pre(a_i) \wedge \nexists a_j \in \mathscr{O} / (a_j, p, a_i) \subseteq CL$*, then* p *is said to be an open condition.*
- $\mathscr{UL}$ *is the set of* unsafe causal links *of* Π*, also called the* threats*. Let* $(a_i, p, a_j) \subseteq \mathscr{CL}$*;* (a_i, p, a_j) *is unsafe if there exists an action* $a_k \in \mathscr{O}$ *such that* $p \in del(a_k)$ *and* $\mathscr{OR} \cup \{a_i \prec a_k \prec a_j\}$ *is consistent.*

Given a planning task $\mathscr{T} = \langle \mathscr{O}, \mathrm{i}, \mathscr{G} \rangle$, a POP algorithm starts with an empty partial plan and keeps refining it until a solution plan is found. The initial empty plan $\Pi_0 = \langle \mathscr{AP}, \mathscr{OR}, \mathscr{CL}, \mathscr{OC}, \mathscr{UL} \rangle$ contains only two dummy actions $\mathscr{AP} = \{a_0, a_f\}$, the *start* action a_0, and the *finish* action a_f, where $pre(a_f) = \mathscr{G}$, $add(a_0) = \mathrm{i}$, $\{a_0 \prec a_f\} \subseteq \mathscr{OR}$, $\mathscr{CL} = \emptyset$, $\mathscr{OC} = \mathscr{G}$ and $\mathscr{UL} = \emptyset$. The empty plan has no causal links or threats, but, has open condition corresponding to the preconditions of a_f (the top-level goals $\mathscr{G}$). A refinement step in a POP algorithm involves two things; first, selecting a flaw (an open condition or a threat) in a partial plan Π, and then selecting a resolver for the flaw. The different ways of solving a flaw are:

- Supporting an open condition with an *action step*. If p is an open condition, an action a needs to be selected that achieves p. a can be a new action from $\mathscr{O}$, or any action that already exists in $\mathscr{AP}$. Solving an open condition involves adding a causal link to Π to record that p is achieved by the chosen action step.
- Solving a threat with an *ordering constraint*. When the flaw chosen is an unsafe causal link (a_i, p, a_j) that is threatened by an action a_k, it can be repaired either by adding the ordering constraint $a_k \prec a_i$, or the constraint $a_j \prec a_k$, into $\mathscr{OR}$. This solving method involves reordering the action steps in Π.

Definition 7. *A plan* $\Pi = \langle \mathscr{AP}, \mathscr{OR}, \mathscr{CL}, \mathscr{OC}, \mathscr{UL} \rangle$ *is **complete** if it has no open conditions* ($\mathscr{OC} = \emptyset$).

Definition 8. *A plan* $\Pi = \langle \mathscr{AP}, \mathscr{OR}, \mathscr{CL}, \mathscr{OC}, \mathscr{UL} \rangle$ *is **conflict-free** if it has no unsafe causal links* ($\mathscr{UL} = \emptyset$).

Definition 9. *A plan* $\Pi = \langle \mathscr{AP}, \mathscr{OR}, \mathscr{CL}, \mathscr{OC}, \mathscr{UL} \rangle$ *is a **solution** if it is complete and conflict-free.*

---

[1] Partial-order planners are capable of handling partially instantiated action instances and hence, the definition of a partial order plan typically includes a set of equality constraints on free variables in $\mathscr{O}$ [9]. We will, however, restrict our attention to ground action instances without any loss of generality for our purposes.

3 Argumentation in POP

The task of the agents in classical planning is to be able to set goals and achieve them, i.e. finding a causal chain of actions that, when applied in the initial state, it achieves the desired (sub)goals. In this sense, the set $pre(a)$ of a planning action a is interpreted as a set of *achievable* preconditions. However, actions can also have preconditions whose predicates are not affected by any of the actions available to the planning agent. Instead, the predicate's truth value is the result of a derivation obtained by forward chaining inference rules. More concretely, in our framework, the agent is equipped with a set of planning actions, $\mathscr{O}$, and a knowledge base $\mathscr{K} = (\Psi, \Delta)$ where:

- Ψ is a consistent set of facts. Initially, $\Psi = \mathrm{i}$, and this set will be updated accordingly with the *add* and *del* effects of the applicable actions.
- Δ is a set of *defeasible* rules that will be used to derive plausible information, tentative conclusions that might be withdrawn with new pieces of information.

In conclusion, a planning action a is a tuple $a = (pre(a), add(a), del(a))$, where the set $pre(a)$ is divided into two subsets:

- $pre\_ach(a)$ denotes the set of *achievable* preconditions of the action a. The semantics is the same as in classical planning; an achievable precondition p of an action a is supported if it exists a set of actions from $\mathscr{O}$ that achieves the fact, and p **holds** in the state in which a will be applied, i.e. p is not deleted by any action before it holds in the state.
- $pre\_der(a)$ denotes the set of *derivable* preconditions of the action a, the set of preconditions that can be solved via a defeasible derivation. More particularly, the semantics is that a derivable precondition p of an action a is supported if there exists an argument $\langle A, p \rangle$ such that the root of a the tree $\mathscr{T}^*_{\langle \mathscr{A}, p \rangle}$ is marked undefeated, i.e. p **is warranted** in the state in which a will be applied.

Achievable preconditions are supported in a partial-order plan through action steps (see section 2.2). On the other hand, derivable preconditions are supported through argument steps as proposed in the argumentation-based formalism presented in [6]. Hence, we define a POP paradigm in combination with the argumentation formalism described in section 2.1, and we analyze the interplay of arguments and actions when constructing plans using POP techniques.

Definition 10. *Let $\mathscr{K} = (\Psi, \Delta)$ be the knowledge base of an agent; and let $\langle \mathscr{A}, p \rangle$, $\mathscr{A} \subseteq \Delta$, an argument that supports a derivable literal p. The set $facts(\mathscr{A})$ contains the facts that appear in the bodies of the rules in $\mathscr{A}$.*

In a partial-order plan Π, when an argument $\langle \mathscr{A}, p \rangle$ is used to support a derivable precondition p of an action a_i, Π will contain a new element, a ***support link*** of the form $(\mathscr{A}, p, a_i)$. This refinement step for solving a derivable precondition of an action is called *argument step* [6]. Like causal links, support links are used to support a derivable precondition with the conclusion of an argument. Assuming an argument step $\mathscr{A}1 = \langle \mathscr{A}, p \rangle$, we can interpret that $add(\mathscr{A}1) = \{p\}$, and $pre(\mathscr{A}1) = facts(\mathscr{A}1)$. As can be observed, the introduction of argument steps does not imply any changes in the POP algorithm.

Under this new perspective, we reformulate the definition 6 as follows: A **partial-order plan** is a tuple $\Pi = \langle \mathscr{A}P \cup \mathscr{A}R, \mathscr{O}R, \mathscr{C}L \cup \mathscr{S}L, \mathscr{O}C \cup \mathscr{D}P, \mathscr{U}L \rangle$, where $\mathscr{A}P$, $\mathscr{O}R$, $\mathscr{C}L$, $\mathscr{O}C$ and $\mathscr{U}L$ have the usual meaning, $\mathscr{A}R$ is the set of argument steps included in Π, $\mathscr{S}L$ is the set of support links, and $\mathscr{D}P$ is the set of pending derivable preconditions of the actions in Π. Note that the facts of an argument step are the achievable preconditions of the argument and as such they are included as open conditions in the set $\mathscr{O}C$.

Unlike the approach presented in [6], DefPlanner is a defeasible argumentation-based planner in which many different agents with different opinions argue with each other on the warranty of a given argument. During the plan construction, at the time of solving a derivable precondition p, DefPlanner invokes a procedure and agents initiate a discussion in order to check whether p can be warranted or not. This procedure builds a dialectical tree for each supporting argument of p and finally returns whether p is defeated or undefeated. This *multi-agent discussion* is explained in detail in next section. Hence, in the case of DefPlanner, argument steps are only inserted in a partial-order plan as long as it has been proven the argument is undefeated. This contrasts with other approaches in which each supporting argument gives rise to a different alternative in the POP algorithm, and discussions on the warranty of a given argument take place in case a counter-argument is introduced in the plan. In conclusion, DefPlanner only inserts provably undefeated arguments in a plan, and, consequently, no threats involving two argument steps may appear in our approach. Let $\langle \mathscr{A}1, p \rangle$ be an argument step inserted in a plan Π; if argument $\langle \mathscr{A}2, q \rangle$ is later inserted in Π then DefPlanner guarantees $\mathscr{A}2$ is not a counter-argument of $\mathscr{A}1$ and viceversa.

Additionally, in this first approach of DefPlanner, we assume a piece of information can not be both derived and achieved. That is, a proposition p is either defeasibly derived through a dialectical tree by using the rules in Δ, or achieved through a course of actions in $\mathscr{O}$. Thus, the predicates of defeasible information are never affected by the available planning actions $\mathscr{O}$ and, consequently, no action-argument threats exist. In section 6, we elaborate on this issue for future versions of DefPlanner.

4 Defeasible Argumentation in a Multi-Agent System

DefPlanner implements a Multi-Agent System (MAS) (figure 1) to assist during the construction plan. Agents can adopt one of the four different roles specified in this MAS:

- *Client role*: The user is represented by an agent playing this role, which is in charge of requesting a plan for a given set of goals.
- *POP role*: The agent playing this role, that is, the planner takes as input the set of goals and returns a solution plan that satisfies the client goals. There is only one agent playing the POP role per MAS.
- *Argumentative role*: An agent ag_i which plays this role is associated with a set of defeasible rules representing the tentative information of the agent about the environment (Δ_i). The task of each argumentative agent ag_i is to participate as far as possible in the multi-agent discussions for warranting a given literal. Each agent has an associated utility function[2] that is used to maximize its benefits.
- *Mediator role*: The agent which plays this role (only one per MAS) is in charge of managing the multi-agent argumentation process.

A MAS, as defined in this paper, is formed by a POP agent which reasons about which action step (for solving an open condition), or ordering constraint (for solving a threat) should be chosen in the next iteration of the POP algorithm; a group of non-self-interested argumentative agents, which join together to reason about the argument step that should be chosen to satisfy/warrant a derivable precondition;

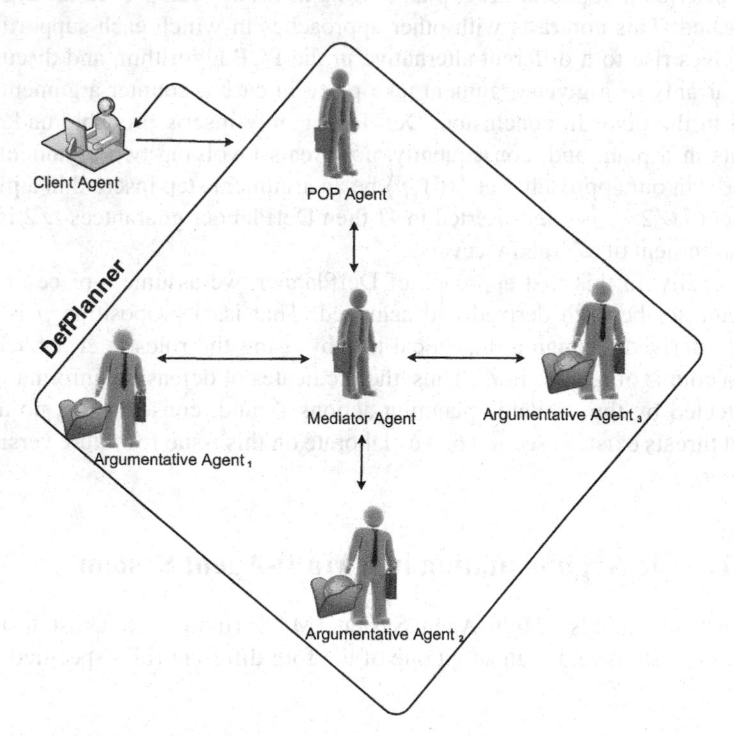

Fig. 1 An overview of DefPlanner.

[2] For instance, in terms of less cost, time, resources or increased safety could be expressed their utility functions.

and, a mediator agent, which coordinates the multi-agent argumentation process for warranting a literal.

4.1 DefPlanner Algorithm

The POP agent implements an extension of the traditional POP algorithm by considering the introduction of argument steps, and corresponding support links, to resolve a defeasible precondition (Algorithm 1). The three non-determinist **choose** statements state that the algorithm has to make a choice among different alternatives (selecting the next partial-plan to work on, selecting a pending derivable precondition in the partial plan, or selecting the next open condition/threat to study). Typically, the selected choice will be the result of the application of a specific heuristic [7]. The *multi_argumentation* function encodes the defeasible argumentation multi-agent process, which will be explained in detail in the next subsection.

The traditional POP algorithm works as follows: starting with the initial empty plan Π_0 (step 1 in Algorithm 1), it works through the application of successive refinement steps at each iteration. First, it chooses a partial-order plan from the list of candidates (step 3 in Algorithm 1), and then it applies a refinement step that involves selecting a flaw (threat or open condition) in the partial-order plan (step 11 in Algorithm 1).

In contrast with the traditional POP algorithm, the new algorithm considers argument steps, besides action steps, to support unsatisfied derivable preconditions. The POP agent takes an argument step as the support from the defeasible argumentation multi-agent process (section 4.2) to derive a defeasible precondition. If no argument steps can be constructed to support a derivable precondition, then it prunes[3] the selected plan Π from *Plan_List*. Note that, unlike the achievable preconditions, the algorithm does not branch for each different argument step that supports a derivable precondition. As it will be explained later, in case of more than one undefeated argument step for a given defeasible precondition, the voting phase will select the best argument step according to the preference criterion of generalized specificity (see section 2.1).

The process ends when both *subgoal_list_1* and *subgoal_list_2* are empty, in whose case Π is a solution plan, or when *Plan_List* is empty, in whose case there is not a solution plan.

4.2 Defeasible Argumentation Multi-Agent Process

The objective of this process is to have multiple agents reasoning (discussing) about the warrant for a particular derivable precondition p requested by the POP agent. The output of the process will be an argument step, if it exists an undefeated argument structure for p; otherwise, the procedure will return *NIL* (step 8 in

[3] i.e. the plan is discarded and the search process does not continue exploring through this plan.

1: $Plan\_list := \Pi_0$
2: **repeat**
3: **choose** $\Pi \in Plan\_list$
$subgoal\_list\_1 := \mathscr{DP}(\Pi)$
$subgoal\_list\_2 := \mathscr{OC}(\Pi) \cup \mathscr{UL}(\Pi)$
4: **if** $(subgoal\_list\_1 \bigcup subgoal\_list\_2 = \emptyset)$ **then**
5: return Π {Plan solution}
6: **else if** $subgoal\_list\_1 \neq \emptyset$ **then**
7: **choose** $\Phi \in subgoal\_list_1$
$\Pi_a := multi\_argumentation(\Phi)$
8: **if** $\Pi_a \neq NIL$ **then**
9: then $Plan\_list := Plan\_list \bigcup \Pi_a$
10: **else**
11: **choose** $\Phi \in subgoal\_list_2$
$Relevant := \{\Pi_r\}, \forall\, \Pi_r$ that resolves Φ {*Each r is a choice (partial-order planning) to solve* Φ}
12: **if** $Relevant \neq \emptyset$ **then**
13: $Plan\_list := Plan\_list \bigcup Relevant$
14: **until** $Plan\_list = \emptyset$
15: return fail {*Not exists plan*}

Algorithm 1. Outline of the DefPlanner algorithm

Algorithm 1) thus indicating there is no refinement plan that supports the defeasible precondition p.

In what follows, we will consider the notions defined in the section 2.1, such as argument structure, disagreement, argumentation line, etc. Unlike single-agent contexts, in our multi-agent framework arguments and counter-arguments will be proposed by different argumentative agents in the MAS.

DefPlanner divides the reasoning process into three phases: the Dialogue Phase, in which arguments and counter arguments are proposed, the Evaluation Phase, in which each argument proposal to derive p is marked as defeated or undefeated, and the Voting Phase, in which a voting is applied - in case of more than one undefeated argument structure- to choose the best undefeated proposal for p according to the preference criterion.

4.2.1 Dialogue Phase

Both the argumentative agents and the mediator agent are involved in this phase. The argumentative agents of the MAS provide two functionalities: (I) propose an initial argument structure to support a derivable precondition p, which will be the root of a dialectical tree, and (II) propose a counterargument to the argument articulated by another agent in the argumentation line. We assume that argumentative agents are ordered according to their indexes: 1, 2, ..., n. The proposed model follows a rotating shift approach[4], in which an argumentative agent can only participate during its turn.

[4] The shift approach allows to treat uniformly each agent.

The mediator agent is in charge of adding the proposed arguments to the appropriate dialectical tree or creating a new dialectical tree in case of a new initial argument structure.

Let $\langle \mathscr{X}^i, h \rangle$ be an argument structure where $\mathscr{X}$ is the argument support, i denotes the argumentative proposer agent, and h is the conclusion supported by the argument. Extending the definition 4 (section 2.1), an argumentation line in DefPlanner, $\Lambda = [\langle \mathscr{X}^i, h_0 \rangle, \langle \mathscr{Y}^j, h_1 \rangle \ldots, \langle, \mathscr{Z}^i, h_n \rangle]$, is a sequence of argument structures from different argumentative agents such that two consecutive argument structures cannot be proposed by the same agent; i.e. $\langle \mathscr{X}^i, h_0 \rangle$, $\langle \mathscr{Y}^j, h_1 \rangle$, and $i \neq j$. Thereby, DefPlanner does not allow agents giving counterarguments to their own arguments, and this is achieved by ensuring that the agent's local belief base (Δ_i) is consistent with respect to the global belief base (Ψ). In this first version of DefPlanner, at the turn of an argumentative agent, it has to articulate all the arguments for a given derivable precondition, or all its counterarguments for a given argument, so an agent can jump-shift the turn only if it lacks sufficient information to make a new proposal. However, in future versions, we will consider to model other different kinds of argumentation strategies.

Specifically, the aim of this phase is to provide reasons that support a derivable precondition $p \in pre\_der(a)$. A new argument $\langle \mathscr{X}^i, p \rangle$ represents the root of a dialectical tree $\mathscr{T}_{\langle \mathscr{X}, p \rangle}$. In order to determine whether $\langle \mathscr{X}^i, p \rangle$ is an undefeated argument or not in the next phase, agents alternatively propose a counter-argument as a defeater to any of the leaf nodes of the dialectical tree $\mathscr{T}_{\langle \mathscr{X}, p \rangle}$. According to [5], a counter-argument $\langle \mathscr{Y}^j, h_2 \rangle$ to the argument $\langle \mathscr{X}^i, h_1 \rangle$ can be a *direct* attack to the conclusion, that is h_2 and h_1 are contradictory literals, or can be an indirect attack by arguing an inner point h of $\langle \mathscr{X}^j, h_1 \rangle$. Since counter-arguments are arguments too, there may exist defeaters for them, and so on, thus giving rise to the argumentation lines of $\mathscr{T}_{\langle \mathscr{X}, p \rangle}$.

4.2.2 Evaluation Phase

At this phase, the aim is to decide whether a dialectical tree of a defeasible precondition p is marked as undefeated or defeated. More specifically, the mediator agent performs a bottom-up-analysis for each dialectical tree $\mathscr{T}_{\langle \mathscr{X}, p \rangle}$ developed in the above phase, obtaining a set of marked dialectical trees $\bigcup_0^n \mathscr{T}^*_{\langle \mathscr{X}, p \rangle}$, where n is the total number of dialectical trees for p. Nodes will be recursively marked as D (*defeated*) or U (*undefeated*) like the minimax tree used in Artificial Intelligence for game trees. At the end of this process, each root argument $\langle \mathscr{X}^i, p \rangle$ will be marked as defeated or undefeated (Definition 5). In DefPlanner, a derivable precondition p is warranted if it has at least a root argument that satisfies p, and the corresponding dialectical tree is marked as U (*undefeated*).

4.2.3 Voting Phase

If the derivable precondition p has more than one undefeated argument, we must choose one of them as the support for p in a partial-order plan Π. In this phase, agents vote the most preferable undefeated argument according to their own utility function. The undefeated argument structure with the highest number of votes will be the selected argument step to be included in Π. In case of tie-breaking, the mediator agent makes the final decision. So, the voting idea is that each agent votes according to their

Different partial plans For instance, the next utility function could be adopted: generalized specificity [14], a function that favors two aspects in an argument to derive a derivable precondition: it prefers (1) a more precise argument (i.e., with greater information content) or (2) a more concise argument (i.e., with less use of rules). So, the undefeated arguments with greater information and less rules would be preferred.

The following section illustrates the application of this protocol to an example scenario in order to obtain a solution plan for a planning task.

5 Example of Application

Figure 2 shows the planning scenario where we will put our argumentation-based model to work. There are two different locations in this scenario $l1$ and $l2$. As can be seen in the figure, there are three different connections between $l1$ and $l2$: via truck, train or plane, and so the client agent can reach $l2$ by using any of these three transport means. The client agent, the truck, the train and the plane are initially located at $l1$. The goal of the problem is to have the client agent in $l2$. Following, we present the objects defined in this problem:

- $l1$, $l2$, ca - location 1, location 2, and the client agent
- tr, tra, pl - a truck, a train, a plane,
- r, tl, al, ae - a road, a railway, an airline company, the airline experts,
- tv, in, - television news, internet news
- bw, sn, wg, - bad weather, snow, wind gusts
- br, vi, ll, esf - bad railroad, adequate visibility, landslides, electrical supply failure
- rm, va, ds - airplane engines work well, volcano ash cloud hits airline, dangerous situation
- h, $j6$, tj - holidays, June 6, and traffic jam.

The actions the client agent can perform are the following ones:

- Mp$(?j, ?k)$: moving plane pl from location j to k. It must exist an airline company to travel from j to k, and absence of dangerous situations to assure safety. Moving a plane takes 3 time units.
- fMt$(?j, ?k)$: fast-moving truck tr from location j to k. It must exist a road from j to k, and assure there is no traffic jam between j and k. This action takes 8 time units.

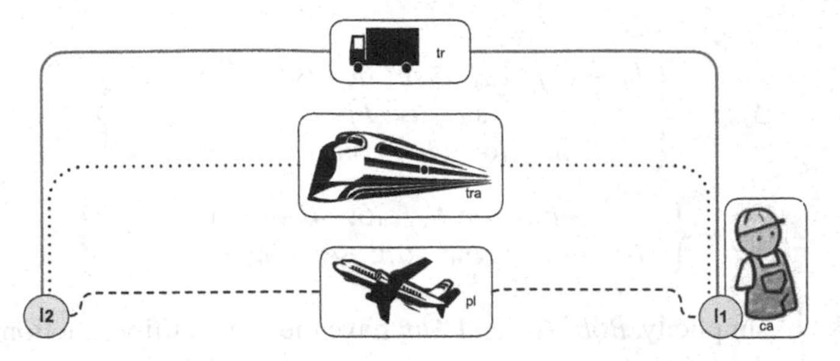

Fig. 2 Scenario of the application example

$$\mathscr{O} = \left\{ \begin{array}{l} \{(at\ tr\ ?k),\ \sim\!\!\{at\ tr\ ?j),\ (at\ ca\ ?k),\ \sim\!\!\{at\ ca\ ?j)\} \xleftarrow{mFt} \\ \quad \{(link\ r\ ?j\ ?k),\ (at\ tr\ ?j),\ (at\ ca\ ?j),\ \sim\!\!\{t\,j)\} \\ \\ \{(at\ tr\ ?k),\ \sim\!\!\{at\ tr\ ?j),\ (at\ ca\ ?k),\ \sim\!\!\{at\ ca\ ?j)\} \xleftarrow{mSt} \\ \quad \{(link\ r\ ?j\ ?k),\ (at\ tr\ ?j),\ (at\ ca\ ?j)\} \\ \\ \{(at\ tra\ ?k),\ \sim\!\!\{at\ tra\ ?j),\ (at\ ca\ ?k),\ \sim\!\!\{at\ ca\ ?j)\} \xleftarrow{mT} \\ \quad \{(link\ tl\ ?j\ ?k),\ (at\ tra\ ?j),\ (at\ ca\ ?j), \sim\!\!\{br),\ (v)\} \\ \\ \{(at\ pl\ ?k),\ \sim\!\!\{at\ pl\ ?j),\ (at\ ca\ ?k),\ \sim\!\!\{at\ ca\ ?j)\} \xleftarrow{mP} \\ \quad \{(link\ al\ ?j\ ?k),\ (at\ pl\ ?j),\ (at\ ca\ ?j),\ \sim\!\!\{ds)\} \end{array} \right\}$$

- sMt($?j, ?k$): slow-moving truck *tr* from location j to k. It must exist a road from j to k. This action takes 20 time units.
- Mt($?j, ?k$): moving train *tra* from location j to k. There must exist a railway from j to k, and no bad railroad conditions to assure an adequate visibility. This action takes 10 time units.

Our multi-agent system consists of the POP agent, the mediator agent and three argumentative agents, *Bob*, *Joe* and *Ann*. Agents have different knowledge and two pieces of information from different agents can appear to be contradictory. Let's assume that each argumentative agent is a travel agency, that *Joe* uses TV as a source of information, but *Ann* prefers Internet to keep up to date. The goal ($\mathscr{G}$) is to have the agent *ca* at position *l*2, (*at ca l*2). The global belief base (Ψ), the local belief bases (Δ_{Bob}, Δ_{Joe}, Δ_{Ann}), and the action base ($\mathscr{O}$) are detailed as follows:

$$\Psi = \left\{ \begin{array}{c} (have\ in);\ (have\ tv);\ (have\ vi);\ (have\ va); \\ (have\ wg);\ (today\ j6);\ (have\ ae);\ (at\ ca\ l1); \\ (at\ tr\ l1);\ (at\ pl\ l1);\ (at\ tra\ l1); \\ (link\ l1\ l2\ r);\ (link\ l1\ l2\ tl);\ (link\ l1\ l2\ al); \end{array} \right\}$$

$$\Delta_{Bob} = \left\{ \begin{array}{c} br \prec ll;\ ll \prec wg;\ bw \prec wg; \\ ds \prec \{va, tv\}; \end{array} \right\}$$

$$\Delta_{Joe} = \left\{ \begin{array}{c} br \multimap esf;\ esf \multimap sn;\ br \multimap sn;\ \sim bw \multimap sn; \\ sn \multimap tv;\ tj \multimap h; \\ h \multimap j6;\ \sim ds \multimap rm;\ rm \multimap ae; \end{array} \right\}$$

$$\Delta_{Ann} = \left\{ \begin{array}{c} \sim bw \multimap h;\ h \multimap j6;\ \sim ll \multimap \sim bw; \\ \sim br \multimap \sim bw; \sim bw \multimap in;\ \sim sn \prec in;\ \sim rm \multimap va; \end{array} \right\}$$

For the sake of simplicity, *Bob*, *Joe* and *Ann* have the same utility function. Specifically, we consider the same comparison criterion among defeaters arguments (section 2.1), as an utility function that returns the best undefeated argument. In what follows, we explain how DefPlanner works to obtain a complete plan Π that satisfies the goal $\mathcal{G}$.

5.1 Searching for a Solution Plan

5.1.1 Step 1

The planning process starts with the empty plan Π_0 (leftmost plan in Figure 5). For solving the precondition $\Phi = (at\ ca\ l2)$, the POP agent has four different action choices $\{mP(l1,\ l2),\ mFt(tr,\ l1,\ l2),\ mT(l1,\ l2),\ mSt(tr,\ l1,\ l2)\}$, so four new partial-order plans $\{\Pi_{0.1}, \Pi_{0.2}, \Pi_{0.3}, \Pi_{0.4}\}$ are added to *Plan_List* (see Figure 6).

5.1.2 Step 2

Let's assume the POP selects the plan $\Pi_{0.1}$ because it is the plan that takes fewer time units. Then we have $\mathcal{AP}(\Pi_{0.1}) = \{mP(l1,\ l2)\}$. The action $mP(l1,\ l2)$ has a derivable precondition $p = \sim ds$ meaning that the plane can only fly if it is assured that no dangerous situation is expected during the flight. The POP agent invokes the mediator agent that calls the *multi_argumentation* function, and it proposes a new dialogue phase to check whether p is warranted or not.

Joe takes the first shift, and puts forward the initial argument $\langle \mathcal{E}^{Joe}, \sim ds \rangle$ with $\mathcal{E}^{Joe} = \{\sim ds \multimap rm;\ rm \multimap ae\}$, indicating that the airline experts assert the airplane engines work well and that there will be no dangerous situation. When counter-arguments to this argument are requested, *Ann* responds[5] with $\langle \mathcal{E}^{Ann}, \sim rm \rangle$ with $\mathcal{E}^{Ann} = \{\sim rm \multimap va\}$, and *Bob* responds $\langle \mathcal{C}^{Bob}, ds \rangle$ with $\mathcal{C}^{Bob} = \{ds \multimap \{va, tv\}\}$. Nobody has more information to argue against, so the process ends here. Figure 3 shows the argument $\langle \mathcal{E}^{Joe}, \sim ds \rangle$ is marked as defeated, and, consequently, $\sim ds$ is not warranted. The *multi_argumentation* function returns $\Pi_{0.1.1} = NIL$ because $\sim ds$ is not warranted. Thereby, $\Pi_{0.1}$ is discarded from the *Plan_list*.

5.1.3 Step 3

Let's assume the next plan to be selected is $\Pi_{0.2}$ (see Figure 6), where $= \mathcal{AP}(\Pi_{0.2}) = \{mFt(tr\_1,\ l1,\ l2)\}$. The action $mFt(tr\_1,\ l1,\ l2)$ has a derivable precondition

[5] An argumentative agent responds if it is at its turn.

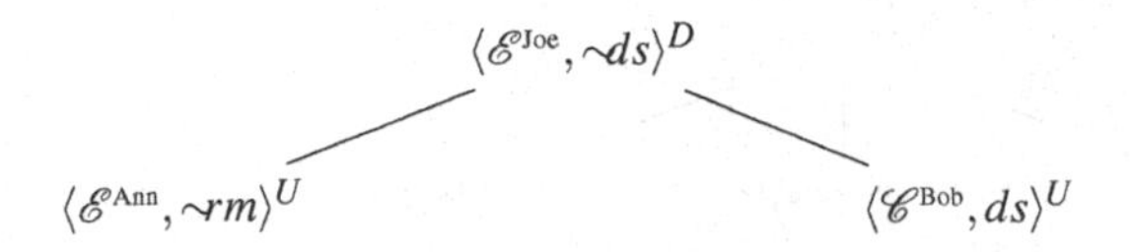

Fig. 3 Marked dialectical tree for the derivable precondition $\sim ds$ at step 2 of the plan solution process

$p = \sim tj$, indicating that there should not be traffic jam for fast-moving truck. The POP agent selects $\Pi_{0.2}$ because it is the second plan with fewer time units. The POP agent invokes the mediator agent that calls the *multi_argumentation* function, and it proposes a new dialogue phase to check whether $\sim tj$ is warranted or not.

Bob and *Ann* have not traffic jam information, and *Joe* knows there is traffic jam because today is June 6, and $\{tj \prec h;\ h \prec j6\}$. Therefore, nobody can support $\sim tj$, and, *multi_argumentation* function returns $\Pi_{0.2.1} = NIL$. Thereby, $\Pi_{0.2}$ is discarded from *Plan_list*.

5.1.4 Step 4

Assuming the next selected plan is $\Pi_{0.3}$ (Figure 6) where $\mathscr{A}P(\Pi_{0.3}) = \{mT(l1,\ l2)\}$, the action $mT(l1,\ l2)$ has a derivable preconditions which indicates the railroad must not be in bad conditions; then, $\Phi = \sim br$.

Ann takes the first shift, and puts forward the initial argument $\langle \mathscr{F}^{\text{Ann}}, \sim br \rangle$ with $\mathscr{F}^{\text{Ann}} = \{\sim br \prec \sim bw; \sim bw \prec in\}$, i.e., internet news say that bad weather is not expected, and, therefore, the railroad will not be in bad conditions. Next, *Bob* takes the shift and responds directly attacking $\sim br$ with $\langle \mathscr{A}^{\text{Bob}}, br \rangle$, where $\mathscr{A}^{\text{Bob}} = \{br \prec ll;\ ll \prec wg\}$, meaning that wind gusts are expected according to the information in the initial state, and because of that landslides may occur. If landslides happen to occur, then it is likely the case to have the railroad in bad conditions.

Joe takes the shift, and responds to $\sim br$ with $\langle \mathscr{B}^{\text{Joe}}, br \rangle$ with $\mathscr{B}^{\text{Joe}} = \{br \prec esf;\ esf \prec sn;\ sn \prec tv\}$, and, $\langle \mathscr{C}^{\text{Joe}}, br \rangle$ with $\mathscr{C}^{\text{Joe}} = \{br \prec sn;\ sn \prec tv\}$. That is, according to Joe's information, television news report it will snow, and so the railroad is likely to be in bad conditions as well as having a electrical supply failure, which causes to have the railroad in bad conditions.

When counterarguments to $\langle \mathscr{B}^{\text{Joe}}, br \rangle$ and $\langle \mathscr{C}^{\text{Joe}}, br \rangle$ are requested, *Ann* responds with $\langle \mathscr{A}^{\text{Ann}}, \sim sn \rangle$, where $\mathscr{A}^{\text{Ann}} = \{\sim sn \prec in\}$. When asked to counter-argue $\langle \mathscr{A}^{\text{Bob}}, br \rangle$, *Ann* responds with $\langle \mathscr{B}^{\text{Ann}}, \sim ll \rangle$ where $\mathscr{B}^{\text{Ann}} = \{\sim ll \prec \sim bw; \sim bw \prec in\}$. According to Ann's information, internet reports that no bad weather is expected and so there is no chance to find landslides.

In turn, when asked to counter-argue $\langle \mathscr{B}^{\text{Ann}}, \sim ll \rangle$, *Bob* takes the shift, and responds $\langle \mathscr{B}^{\text{Bob}}, bw \rangle$ with $\mathscr{B}^{\text{Bob}} = \{bw \prec wg\}$. In turn, *Joe* responds $\langle \mathscr{D}^{\text{Joe}}, \sim bw \rangle$ with $\mathscr{D}^{\text{Joe}} = \{\sim bw \prec sn;\ sn \prec tv\}$, and, *Ann* responds $\langle \mathscr{C}^{\text{Ann}}, \sim bw \rangle$ with $\mathscr{C}^{\text{Ann}} = \{\sim bw \prec h;\ h \prec j6\}$, and $\langle \mathscr{D}^{\text{Ann}}, \sim bw \rangle$ with $\mathscr{D}^{\text{Ann}} = \{\sim bw \prec in\}$.

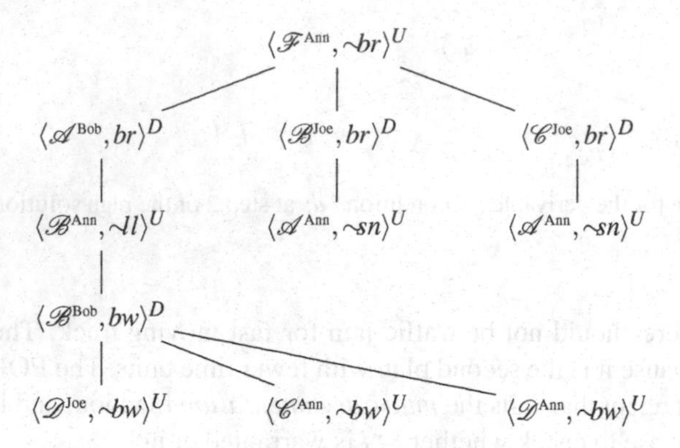

Fig. 4 Marked dialectical tree for the derivable precondition $\sim br$ at step 2 of the plan solution process

Figure 4 shows that the argument $\langle \mathscr{F}^{\text{Ann}}, \sim br \rangle$ is marked as undefeated, and, consequently, the derivable precondition $\sim br$ is warranted. The *multi_argumentation* function returns $\Pi_{0.3.1}$, an extension of $\Pi_{0.3}$ with $\mathscr{F}^{\text{Ann}}$ and $CL(\mathscr{F}^{\text{Ann}})$.

5.1.5 Step 5

Assuming the plan selected next is $\Pi_{0.3.1}$ (because it has less duration than $\Pi_{0.4}$), the POP agent extends $\Pi_{0.3.1}$ to $\Pi_{0.3.1.1}$, adding a causal link between $facts(\mathscr{F}^{\text{Ann}})$ and the initial state Ψ (Figure 6). $\Pi_{0.3.1.1}$ is a solution plan that satisfies the goal $\mathscr{G}$ (Figure 5).

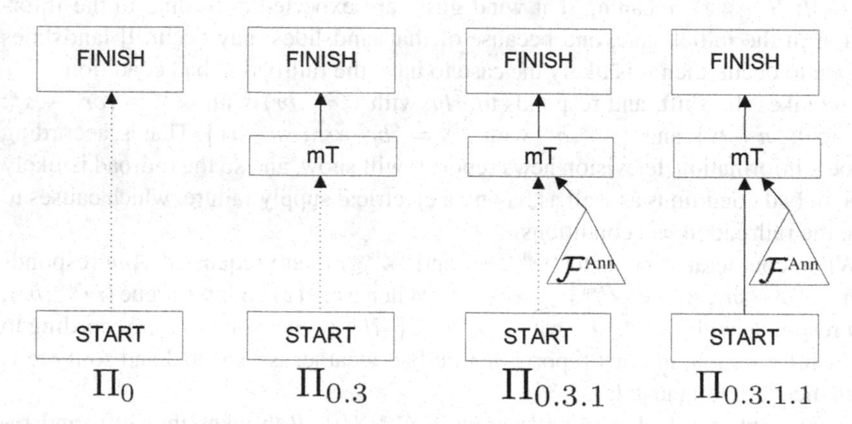

Fig. 5 Different partial plans for the example scenario

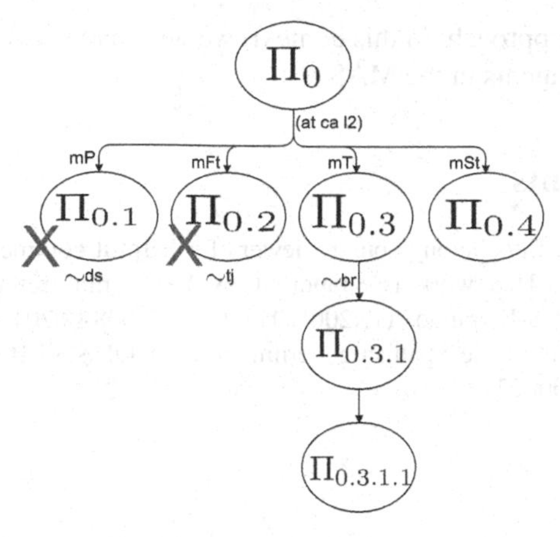

Fig. 6 Search in the space of partial-order plans for the example scenario

6 Conclusions and Related Work

In this paper, we have presented DefPlanner, a defeasible argumentation-based planner that allows multiple agents with partial and contradictory knowledge articulate reasons for and against the precondition of a planning action. Along the paper, we have introduced the necessary modifications to include a defeasible reasoning into a POP algorithm. This new and enriched planner opens up many possibilities to be applied to a multi-agent planning context.

DefPlanner builds on the approximation of Garcia et al [6], and extends their work by incorporating multiple agents at the time of deciding which literals (conditions of a planning action or derivable preconditions) are warranted. Our work is also related to conformant planning [8], an approach to deal with planning with incomplete information in which the purpose is to generate plans given uncertainty about the initial state and action effects, and without any sensing capabilities during plan execution. However, unlike conformant planning, our approach is a powerful planning mechanism for reasoning about contradictory information coming from different sources or agents. In this sense, in the literature of classical planning we can hardly find approaches to deal with contradictory information because, among other reasons, there are very few attempts to extend planning to a multiagent environment, being a notably exception the work of Brenner and Nebel [3]. Hence, DefPlanner is a novel approach regarding the consideration of incomplete and contradictory information of multiple reasoning entities, i.e. agents.

As for future work, we are interested in extending the argumentation process to achievable preconditions; that is, a new approach towards the integration of reasoning about action steps (practical reasoning) and reasoning about argument steps (defeasible reasoning). Particularly, our next immediate step is to endow agents with planning capabilities, rather than just limiting agents to perform defeasible reasoning and discuss the warranty of literals, and thus come up with a defeasible multiagent planning approach. In this context, we will also study the choice of having non-cooperative agents in the MAS.

Acknowledgements

We would like to thank three anonymous reviewers for helpful comments that have helped to improve this work. This work is supported by FPU grant reference AP2009-1896 awarded to Sergio Pajares-Ferrando, TIN2008-04446, TIN2008-06701-C03-03 and PROMETEO/2008/051 projects of the Spanish government and CONSO-LIDER INGENIO 2010 under grant CSD2007-00022.

References

1. Barrett, A., Weld, D.S.: Partial-order planning: evaluating possible efficiency gains. Artificial Intelligence 67(1), 71–112 (1994)
2. Bench-Capon, T.J.M., Dunne, P.E.: Argumentation in artificial intelligence. Artificial Intelligence 171(10-15), 619–641 (2007)
3. Brenner, M., Nebel, B.: Continual planning and acting in dynamic multiagent environments. Journal of Autonomous Agents and Multiagent Systems 19(3), 297–331 (2009)
4. Fikes, R.E., Nilsson, N.J.: STRIPS: A new approach to the application of theorem proving to problem solving. Artificial intelligence 2(3-4), 189–208 (1971)
5. García, A.J., Simari, G.R.: Defeasible logic programming: An argumentative approach. Theory and Practice of Logic Programming 4(1-2), 95–138 (2004)
6. Garcıa, D.R., Garcıa, A.J., Simari, G.R.: Defeasible Reasoning and Partial Order Planning. In: Hartmann, S., Kern-Isberner, G. (eds.) FoIKS 2008. LNCS, vol. 4932, p. 311. Springer, Heidelberg (2008)
7. Ghallab, M., Nau, D., Traverso, P.: Automated Planning. Theory and Practice. Morgan Kaufmann, San Francisco (2004)
8. Hoffmann, J., Brafman, R.I.: Conformant planning via heuristic forward search: A new approach. Artif. Intell. 170(6-7), 507–541 (2006)
9. Penberthy, J.S., Weld, D.: UCPOP: A sound, complete, partial order planner for ADL. In: Proceedings of the Third International Conference on Knowledge Representation and Reasoning, pp. 103–114. Citeseer (1992)
10. Prakken, H., Gordon, T., Walton, D., Bench-Capon, T., Bex, F.J., den Braak, S.W.v, Oostendorp, H.v., Prakken, H., Verheij, H.B., Vreeswijk, G.A.W.: Logical tools for modelling legal argument: a study of defeasible reasoning in law, Dordrecht, Boston (1997)
11. Prakken, H., Sartor, G.: Argument-based extended logic programming with defeasible priorities. Journal of Applied Non-Classical Logics 7(1) (1997)
12. Rahwan, I., Amgoud, L.: An argumentation-based approach for practical reasoning. In: Maudet, N., Parsons, S., Rahwan, I. (eds.) ArgMAS 2006. LNCS (LNAI), vol. 4766, pp. 74–90. Springer, Heidelberg (2007)
13. Simari, G.R., García, A.J., Capobianco, M.: Actions, planning and defeasible reasoning. In: 10th International Workshop on Non-Monotonic Reasoning (NMR 2004), pp. 377–384 (2004)

14. Stolzenburg, F., Garcia, A.J., Chesnevar, C.I., Simari, G.R.: Computing generalized specificity. Journal of Applied Non-Classical Logics 13(1), 87 (2003)
15. Thimm, M.: Realizing argumentation in multi-agent systems using defeasible logic programming. In: McBurney, P., Rahwan, I., Parsons, S., Maudet, N. (eds.) ArgMAS 2009. LNCS, vol. 6057, pp. 175–194. Springer, Heidelberg (2010)
16. Thimm, M., Kern-Isberner, G.: A distributed argumentation framework using defeasible logic programming. In: International Conference on Computational Models of Argument (COMMA), pp. 381–392 (2008)
17. Weld, D.S.: An introduction to least commitment planning. AI magazine 15(4), 27 (1994)

Operator Behavior Modelling in a Submarine

Isabelle Toulgoat, Pierre Siegel, and Yves Lacroix

Abstract. Simulations of naval action estimate the operational performance of warships or submarines for a given scenario. In common models, the operator's reactions are predefined. This is not realistic: the operator's decision can produce unexpected reactions.

This paper presents a method to model operator decision in simulations. This method allows to reason about incomplete, revisable and uncertain information: an operator has partial information about his environment only and must revise his decisions. Our method uses a nonmonotonic logic: the rules of behavior are formalized with default logic, to which we added a consideration of time. Our method uses preferences to manage choice between different rules, with simple probabilistic techniques.

This method has been implemented in Prolog, interfaced to DCNS simulator framework and applied to a scenario involving two adverse submarines.

1 Introduction

Simulations of naval action estimate the operational performance of warships or submarines for a given scenario. For example, a submarine must be discreet, in order not to be detected by an ennemy. One of the main aspect regarding

Isabelle Toulgoat
DCNS Ingénierie, Le Mourillon, BP 1306, 83 076 Toulon Cedex, France
e-mail: isabelle.toulgoat@dcnsgroup.com

Pierre Siegel
Université de Provence, France
e-mail: siegel@cmi.univ-mrs.fr

Yves Lacroix
Systèmes Navals Complexes, Avenue Georges Pompidou, 3160, La Valette du Var, France
e-mail: yves.lacroix@univ-tln.fr

I. Hatzilygeroudis and J. Prentzas (Eds.): Comb. of Intell. Methods and Appl., SIST 8, pp. 21–39.
springerlink.com

operational performance of a submarine is the detection acoustique performance and the risk of being detected by an ennemy. At the DCNS, the simulator framework ATANOR models complex scenarios involving several platforms with combat system and equipment [18].

In this simulator framework, the behavior is modelled with Petri nets [13], composed of places, which model the equipment states and transitions between these places. These transitions are activated by internal and external events. Only one place is activated at the same time, which forbids simultaneous actions.

The modelling of behavior rules with Petri nets provides automatic reactions of the combat system to a tactical situation. This is not realistic: in a tactical situation the decision of an operator is a key aspect, which can provide unexpected reactions. Moreover, a disadvantage of the modelling with Petri nets is to have to revise its implementation for any new behavior [7].

The purpose of this work is to develop a system allowing to model the behavior of an operator in the performance simulations. We worked on a case study involving two adverse submarines. This system has to obey several requirements:

- to be able to model the behavior rules of the operator.
- to be able to reason with incomplete, revisable and uncertain information. Indeed, an operator has a partial sight of this environment only. This environment is always changing: the submarine can lose the detection, it hasn't the exact position of its adversary, it is just an estimation.... Therefore he must reason with uncertain and incomplete information. His decisions must be revised with the arrival of new information [17] [6].
- to choose between different proposals when the system proposes several actions for a same situation.
- to allow the addition of new behavior rules, without having to modify the knowledge representation and without calling into question the previous rules (unlike the Petri nets, in which the modifications are complicated).
- to be able to reason with general rules, without having to compile in a very precise way all the information. It is not necessary for the user to describe all the possibilities.

This work is financed by the DCNS company for military applications: we need a simple and robust program. Therefore, we used the most widely known nonmonotonic logic: the default logic. We added a consideration of time: we have submarine's data at the time t, and the extensions calculus gives all the possible extensions at the next time $t + 1$. Each extension is a proposal for the action of the submarine. We calculate a weight function for each extension, thanks to preferences on defaults. Then, we use simple probabilistic techniques to choose between these extensions. This work has been implemented using Prolog, and interfaced with DCNS simulator [21], [19].

In the following paper we will first present the case study and some behavior rules. Then, we will present the limits of the classical logic and why we need the nonmonotonic logic. We will explain the formalization of the behavior rules with default logic. We use only normal defaults and Horn clauses in order to simplify the program, though we could extend this work to other case studies, with more complicated rules. Next we explain the choice between the extensions thanks to preferences with simple probabilistic techniques. Finally, some results are presented.

2 Case Study: Submarine Detection and Tracking

In a scenario including two submarines, we model the decision of an on-watch officer in the submarine according to the events perceived on the tactical situation. In this purpose, we questioned submariners about this case study, and we inferred behavior rules. These rules are really used in a submarine. During this work, we always got in touch with submariners, in order to complete the rules.

Here are some examples of these rules:

- Rule 1: As long as the submarine has no detection, it continues a random research trajectory in its patrol area. During that process, the submarine makes successive straight sections: it goes straight ahead and sometimes it changes its course. The submarine is deaf in its rear (behind the submarine, the sonar's reception is decreased for several reasons), this manœuver allows the submarine to check that it isn't tracked. With this manœuver, the submarine covers the entire patrol zone, in order to increase its chances to detect an intruder.
  Remark: it is a rule of minimal change [8] [23] [6]. This rule is applied as long as the submarine has no new information.
- Rule 2: If the submarine detects another submarine, the officer engages the following actions:
  - Collision avoidance manœuver.
  - Elaboration of the solution manœuver: he manœuvers in order to confirm his information about the distance, the course and the speed of the enemy.
  - Bypassing of the enemy manœuver: when the officer is sure not to be detected, he gets closer to the enemy's rear, position in which he won't be detected.
  - Tracking manœuver: when the submarine is in the enemy's rear, it begins the tracking: it makes straight sections in the enemy's rear, avoiding to be detected and keeping good information about the enemy's kinematics.
- Rule 3: If the submarine is detected when it makes one of the following manœuvers: elaboration of the solution, bypassing of the enemy or tracking, it must escape: the officer manœuvers in order to go away from the enemy, aiming the loss of contact.
- Rule 4: If the submarine is a diesel submarine and more than a few hours have passed since the last battery charge, it must rise to the surface and use the snorkel to take air from the surface and evacuate exhaust gas.
- Rule 5: If the submarine loses the contact during the tracking, the officer rallies the last position of the adversary and searches for it.

If he finds it, then he resumes the tracking actions (Rule 2).
If after one hour he hasn't found it, he resumes the random research trajectory (Rule 1).
During this research hour, the submarine can not rise and use the snorkel (Rule 4).

- Rule 6: With the sonar called MOAS (Mine and Obstacle Avoidance System), the submarine can detect mines, big rocks, cliffs. If the submarine detects a big rock, it changes its course in order to place the rock to one side.

These rules can be in competition: at a same time, it is possible that the submarine needs to do several actions. For example, it needs to rise the surface and use the snorkel and it needs to continue the tracking. The system must be able to manage these alternative choices.

3 Classical Logic and Its Limits

The classical logics, as the mathematical or the propositional logics, are monotonic: if we add information or a formula E' to a formula E, everything which was deduced from E will be deduced from E ∪ E'.This monotonicity will generate problems to reason with incomplete, uncertain and revisable information. Indeed, in this case, it can happen that previously established conclusions turn invalid due to new information arrival or information change.

- The classical logic doesn't allow to reason about incomplete information. Let us take the rule: "Generally, a submarine with no detection makes a random research trajectory". At first sight, we can express this type of information within the first-order logic:
  Rule 1:
  $$\forall x, \neg detection(x) \rightarrow random\_trajectory(x) \quad (1)$$
  This formulation is coherent if the only known information is "The submarine has no detection".
  But if we had the rule: "If more than four hours have passed since the last battery charge, the submarine must rise to the surface and use the snorkel to take air.", we express it within first-order logic:
  Rule 2:
  $$\forall x, Tlc(x) \geq 4 \rightarrow snorkel(x) \quad (2)$$
  where Tlc denotes the time since the last charge and $snorkel$ the action of rising to the surface and using the snorkel.
  With these rules, it is difficult to manage general rules containing an important number of exceptions [17].
- The classical logic doesn't allow to revise the information: it doesn't plan to revise the previously established deductions. Let us take again the rules 1 and 2. Knowing that the submarine has no detection, we deduce that it must make a

random trajectory. But, if we know that more than four hours passed since the last battery charge, we conclude that the submarine must use the snorkel.
We obtain two conclusions which are not consistent: the submarine can not make at the same time these two actions.
It illustrates how classical logics don't allow revising the reasoning and the conclusions. This kind of reasoning is common in artificial intelligence, as well as in the daily life.

In the case of a submarine, blind in submersion, the only information comes from passive sonar system, this information is uncertain and incomplete [15], [3]. The officer must be able to revise the decisions with the arrival of new information. We need a logic which allows to reason about incomplete, uncertain and revisable information.

4 Nonmonotonic Logic and Default Logic

A nonmonotonic logic allows to eliminate the monotony property of the classical logic: if a reasoning gives some conclusions using some given knowledge, these conclusions could be revised with the addition of new knowledge.

A nonmonotonic logic allows to take the incomplete, revisable, uncertain information into account. This logic has a natural similarity with the human reasoning: due to the lack of information or lack of time, one can reason with partial knowledge and revise the conclusions when one has more information.

The default logic, introduced by Ray Reiter [16], is the most widely used logic. It formalizes the default reasoning: conclusions can be made, in the absence of opposite proof. A default logic is defined by $\Delta = (D, W)$, W is a set of facts (formulae from propositional logic or the first-order logic), and with D, a set of defaults, (inference rules with specific content, which handle uncertainty).

Let us remind the definitions of defaults and extensions:

Definition: Default

A default is an expression of the form:

$$\frac{A(X) : B(X)}{C(X)} \tag{3}$$

where A(X), B(X) and C(X) are formulae and X is a set of variables. A(X) is the prerequisite, B(X) is the justification and C(X) is the consequent. Intuitively, this default (formula 3) means: if A(X) is true, if it is possible that B(X) is true (B(X) is consistent), then C(X) is true.

If $B(X) = C(X)$, the default is normal. The normal default means: " Normally, the As are Bs".

Definition: Extension

The use of defaults allows to deduce more formulae from a knowledge base W. To generate the deduced formulae, we calculate extensions, which are defined as follows:

E is an extension of Δ if and only if $E = \cup_{i=0,\infty} E_i$, with

$$E_0 = W \; and \; for \; i \geq O,$$

$$E_{i+1} = Th(E_i) \cup \{C/(\frac{A:B}{C}) \in D, A \in E_i, \neg B \notin E\}$$

where $Th(E_i)$ is the set of theorems obtained in a monotonic way from E_i.

It is important to notice that E appears in the definition of E_{i+1}. So, we need to know E to find E_i, it is not possible to obtain the extensions with an incremental algorithm.

If we work with normal defaults, the definition of an extension is changed: we need to verify that $\neg B \notin E_i$:

$$E_0 = W \; and \; for \; i \geq O,$$

$$E_{i+1} = Th(E_i) \cup \{B/(\frac{A:B}{B}) \in D, A \in E_i, \neg B \notin E_i\}$$

where $Th(E_i)$ is the set of theorem obtained in a monotonic way from E_i.

For our case study, we only use normal defaults, but we could extend our work to general defaults.

5 Rules Formalization with Default Logic

5.1 *Time Consideration*

To formalize the rules of behavior, we used default logic, to which we added a consideration of time. Indeed, we have submarines data at the time t, and we have to deduce the submarines instructions at the next time t+1, taking into account the state of the submarine and updated information. These instructions will generate the submarine updated data for time t+1. To introduce the time, we used previous work by Cordier and Siegel [6].

We need the time consideration in the definitions of the facts W and the defaults D of the default logic $\Delta = (D, W)$.

5.2 *Facts Definition with Time Consideration*

The set of facts W is defined with formulae from propositional logic or first-order logic.

- We used basic facts (or ground literals) like, for example: $detection(d_t)$, $course(c_t)$, $speed(s_t)$... The basic facts define the submarine information at the time t.
- We use only Horn clauses. They allow us to write two types of rules:
  - the Horn clauses with a positive literal, written as follows:

$$(g(t) \vee \neg f_1(t) \vee \ldots \vee \neg f_k(t)) \tag{4}$$

    where the $f_i(t)$ and $g(t)$ are positive literal at time t. This formula can also be written with an implication:

$$(f_1(t) \wedge \ldots \wedge f_k(t)) \rightarrow g(t) \tag{5}$$

    This type of rules allows to define rules which are always true, these are classic rules of expert systems.
    Example: we formalise a rule such as "If the submarine has a random research trajectory, it turns by an angle between α and β", as follows:

$$random\_trajectory(X_t) \rightarrow turn(X_t, (\alpha, \beta)) \tag{6}$$

  - the Horn clauses with no positive literal, written as:

$$(\neg f_1(t) \vee \ldots \vee \neg f_k(t)), \tag{7}$$

    ie

$$\neg(f_1(t) \wedge \ldots \wedge f_k(t)) \tag{8}$$

    We use these rules to define mutual exclusions in pairs, these are the predicates which can not be executed at the same time:

$$(\neg f_1(t) \vee \neg f_2(t)) \tag{9}$$

    equivalent to

$$\neg(f_1(t) \wedge f_2(t)) \tag{10}$$

    Example: we can define a rule such as "The submarine can not make at the same time a random research trajectory and rise to use the snorkel" as follows:

$$\neg(random\_trajectory(X_t) \wedge snorkel(X_t)) \tag{11}$$

The basic facts and the Horn clauses are easily understandable for the users, and we obtain a simple program. However, we could generalize our work to other clauses.

5.3 Default Definition with Time Consideration

The defaults D are inference rules with specific content, they allow to manage uncertainty. They express the fact that, if there is no contradiction to execute an action, the submarine can do it. We use here only normal defaults. The normal defaults allow to ensure the existence of at least one extension [16] and to obtain a simple algorithm of extension calculus. However, we could generalize our work with other defaults.

They allow us to formalize rules such as "If the submarine has no detection, then it makes a random research trajectory" in the following way:

$$\frac{\neg detection(X_t) : random\_trajectory(X_{t+1})}{random\_trajectory(X_{t+1})} \tag{12}$$

This default means: "If the submarine has no detection at time t and if it can make a random research trajectory at time t+1, it makes a random research trajectory at time t+1".

The defaults allow us to define general rules on the behavior of the submarine (rise to the surface to use snorkel, collision avoidance, tracking...). Then, the set of facts allows to specify, for each behavior, the action to realize (change course, speed, submersion) and the mutual exclusions between the behaviors.

5.4 Extension Calculus

We use extensions calculus to study all the defaults and to retain the defaults that answer the problem in a coherent way. Each extension is a possible solution to the problem: according to the submarine state at the time t, an extension gives a possible solution of action for the submarine at the next time t+1 [20]. The normal defaults grant the existence of at least one extension. Generally, we will have several extensions for the same knowledge base.

We could use the answer set programming [12] to calculate the extensions, which are equivalent. In order to have a simple system, we rather implement our own extension calculus. The normal defaults and the Horn clauses allow to implement easily the extensions calculus with the Prolog language. We called our program NoMROD for Non-Monotonic Reasoning for Operator Decision.

6 Extensions Selection with Preferences

The aim of this part is to simulate the officer decision. In a tactical situation, the decision of an operator is a key aspect. At each time, the officer has to choose between the actions. We define a method to choose between the different extensions proposed. This method allows to simulate different types of behavior, depending on the officer's character.

We distinguish two stages in this method. The first stage consists in defining general principles for the extension selection, and a weight function for the extension with a multi-criteria decision aid method. This weight function is a general formula, which can be used for other case studies. This function handles the officer's behavior at two levels:

- the importance of each default according to different criteria, with preference coefficients $C_1, \ldots, C_n$, which allow to define preferences on defaults.These coefficients describe the importance of each action.
- the officer character with the character coefficients $\beta_1, \ldots, \beta_n$: which criteria will he favor? Thanks to these character coefficients, we can model different officers: careful, bold...

Next, the second stage is more specific to the case study. We define a method to select an extension, thanks to the weight function.

6.1 Extension Selection Principles

First, we study some principles and requirements to which the extension selection must answer:

1. to choose the interesting extension and let the others aside.
   Example: NoMROD proposes two extensions: to make a random research trajectory or to rise to the surface and use the snorkel to take air.
   The random research trajectory is a rule of minimal change: it is applied while the submarine has no new information. The officer will choose to rise to the surface, which is a more important behavior.
2. to choose the extensions which are obligatory (for the crew survival).
   For example, the officer can postpone the rising to use the snorkel, if he is doing another important action (for example the tracking). When this rule is verified, the officer has approximately thirty minutes of battery. When this reserve will be practically empty, the officer will be obliged to rise.
3. to manage the choice between several extensions.
   Example:NoMROD proposes two extensions: avoid the collision with a submarine and avoid the collision with a big rock. These two behaviors are very important for the submarine safety. NoMROD must be able to choose between extensions which have the same importance.
4. to respect the minimal change: while the submarine has no new information, it stays in the same state, it doesn't change its behavior. We must give more chance to an action already engaged.
   With this rule, the officer will persist in its choices, he won't oscillate between several behaviors. However, NoMROD must be able to stop an action if another becomes obligatory.
   Example: NoMROD proposes two extensions: to track the enemy and to rise to the surface and use the snorkel to take air. We suppose that the system chooses the tracking. These two extensions will be proposed again while the rising won't

be executed. The best choice for the officer is to carry on the tracking, and when the rising becomes obligatory (the submarine has just enough battery to rise to the surface), he must execute this action.
When the submarine begins an action, it seems important to carry on this action during a certain time. The officer can not change his opinion too often.
5. the enemy submarine doesn't have to guess which action our officer will select.

These principles are some principles of common sense. Aside from last principle, the other can be applied generally to other case studies.

6.2 Extensions Weight Function

We define a weight function to give weights to the extensions.These weights quantify the extensions importance. In order to define this weight function, we use a method of multi-criteria decision aid (MCDA). MCDA aims at modelling the preferences of a decision maker. It allows the decision maker to solve complex problems, where several criteria must be handled in the choice [2],[22]. There are several categories of methods in MCDA. We use the multi-attribute utility theory. This theory is based on the following axiom:

Every decision maker tries unconsciously to maximize a function

$$U = U(g_1, \ldots, g_n), \tag{13}$$

which aggregates all the points of view to be handled (with g_i the criteria).

In order to define a general method to define the extensions weight functions, we use a simple aggregation function: the weighted sum.

Each extension uses defaults. First, we attribute preference coefficients on defaults. Next, we calculate the utility function for each extension, and finally we calculate the weight functions for each extension using the weighted sums.

6.2.1 Preference's Coefficients on Defaults

Each default is a general behavior. In order to specify the importance of behaviors, the user allocates preference coefficients to the defaults.

In a similar way, the preferences are used within a system of nonmonotonic reasoning, allowing finding an appropriate compromise solution. For example, Brewka [4] defined a prioritized default logic, with a definition of order in which defaults must be applied.

Different preference coefficients $C_1, \ldots, C_n$ are attributed to specify the importance of the defaults according to different criteria.

For example, we can specify the default importance for the submarine safety, the efficiency on the submarine mission, the order obedience, ecologist criterion...
For each default D_j, we attribute values to these coefficients $C_{1j} \ldots C_{kj}, \ldots, C_{nj}$.
We fixed arbitrarily these coefficients between 0 and 1000.

6.2.2 Utility Function

Each extension E_i uses defaults:

$$E_i = \{D_1 \ldots, D_m\}$$

For each extension E_i, we calculate the scores

$$\{Score_1(E_i), \ldots, Score_n(E_i)\}$$

which correspond respectively with the coefficients $C_1, \ldots, C_n$.

A score $Score_k(E_i)$ is the sum of the coefficients C_k of each default used in the extension E_i:

$$Score_k(E_i) = \sum_{j=1}^{m} C_{kj} \tag{14}$$

We suppose that NoMROD proposes p extensions. For each $Score_k(E_i)$ for an extension E_i, we calculate a utility function $\mu_k(E_i)$. The sum of the utility functions μ_k must be equal to 1:

$$\sum_{j=1}^{p} \mu_k(E_j) = 1 \tag{15}$$

The score $Score_k(E_i)$ is divided by the sum of all the $Score_k(E_j)$ of the p extensions proposed by the system:

$$\mu_k(E_i) = \frac{Score_k(E_i)}{\sum_{j=1}^{p} Score_k(E_j)} \tag{16}$$

$\mu_k(E_i)$ is the evaluation of the extension E_i, according to the criterion C_i.

6.2.3 Officer's Character

We want to model the officer's character. According to his character, the officer will use different tactics.

For example, a careful officer will give more importance to the behaviors which ensure the submarine safety. A bold officer will favor the efficient behavior for the submarine mission. To model these characters, the user defines character coefficients $\beta_1, \ldots, \beta_n$ in the weight function, such as the sum of these coefficients is equal to one

$$\sum_{k=1}^{n} \beta_k = 1 \tag{17}$$

6.2.4 Weight Function

Finally, we use a simple function to aggregate criteria: the weighted sum. For each extension E_i, we have:

$$P(E_i) = \sum_{k=1}^{n} \beta_k \mu_k(E_i) \tag{18}$$

with the equation 17. The sum of the extension weight functions has the following property:

$$\sum_{i=1}^{p} P(E_i) = 1. \tag{19}$$

We obtain a general formula, which could be easily reused with other case study.

The use of the weighted sum as aggregation function implies some limitations [10]. More specially, this function can favor the extreme extensions (for example, an extension with a utility function very small for a criterion and with a utility function very important for an other criterion) to the detriment of an other extension with more well-balanced utility functions. Other aggregation functions, such as the Choquet intergal [9], allow to solve this problem by handling the interaction between criteria. We plan to test this aggregation function in future work.

6.2.5 Example of Weight Function Calculus

We define four defaults:

$$\{D_1, D_2, D_3, D_4\},$$

and two coefficents: the submarine safety C_{safety}, and the efficiency on the submarine mission $C_{efficiency}$.

We attribute values to each default:

Defaults	C_{safety}	$C_{efficiency}$
D_1	500	600
D_2	10	800
D_3	1000	900
D_4	50	60

We suppose we have to choose between two extensions:
$E_1 = \{D_1, D_4\}$ and $E_2 = \{D_2, D_3, D_4\}$.

We calculate the extensions scores (cf. formula 14):

Scores	E_1	E_2
$Score_{safety}$	550	1060
$Score_{efficiency}$	660	1760

We calculate the utility functions (cf. formula 16):

μ	E_1	E_2
μ_{safety}	$\frac{550}{1610}$	$\frac{1060}{1610}$
$\mu_{efficiency}$	$\frac{660}{2420}$	$\frac{1760}{2420}$

Finally, we have the weight functions (cf. formula 18):

$$P(E_1) = \beta_{safety} * \mu_{safety}(E_1) + \beta_{efficiency} * \mu_{efficiency}(E_1)$$

$$P(E_2) = \beta_{safety} * \mu_{safety}(E_2) + \beta_{efficiency} * \mu_{efficiency}(E_2)$$

If the officer prefers to favor the safety of his submarine rather than the mission efficiency, we can define: $\beta_{safety} = 0.6$ and $\beta_{efficiency} = 0.4$. With this example, we obtain:

$P(E_1) = 0.31$ and $P(E_2) = 0.69$.

6.3 Random Extension Choice

In a tactical situation, the decision of an operator is a key aspect, which can provide unexpected reactions. In our case study, the choice hasn't to be always determinist. Moreover, with deterministic reactions, it is easy for the opposing submarine to guess these reactions (principle 5). We need a choice method, which handles these unexpected reactions.

For these reasons, we don't choose always the extension with the maximum weight function. We prefer to introduce an additional part of uncertainty with a random choice. However, this random choice must be coherent with the principles which guide the officer's decision, and with the extension selection principles defined previously.

6.3.1 Random Choice

The random choice is based on a random sampling: we have p extensions:

$$E_1, \ldots, E_p$$

and the respective weight functions:

$$P(E_1), \ldots, P(E_p)$$

with the formula 19. Each weight function $P(E_i)$ is the probability for the extension to be chosen.

Example 1: We have to choose between two extensions:

- E_1 with the weight function $P(E_1) = 0.1$.
- E_2 with the weight function $P(E_2) = 0.9$.

With this random choice, we manage the choice between several extensions (principle 3).

6.3.2 Correction on the Extension Weight Function

With this random sampling, we have problems to solve. The random choice is realist if we have to choose between extensions with close weight.

However, if we have an extension with a very important weight, we want to choose this one (example 1).

We have the same problem in the following example (example 2). The system proposes six extensions: five extensions with the same weight function 0.1 and one with the weight function equal to 0.5. The random sampling gives as much chances to be chosen to the five extensions with the weight function equal to 0.1: $5 * 0.1 = 0.5$, as to the extension with the weight function equal to 0.5. In such a case, it seems more natural to choose the extension with the weight function equal to 0.5.

We have to modify the weight function, in order to give a more important weight to the important extension, and less important weights to the others. In this purpose, we apply a correction to the weight function: the power function $f(x) = x^k$, with $k > 1$.

The correction is applied as follows:

- We have to choose between p extensions :
$$E_1, \ldots, E_p,$$
and the respective weight functions:
$$P(E_1), \ldots, P(E_p),$$
with the formula 19.
- We apply the power function
$$f(x) = x^k \tag{20}$$
with $k > 1$: $P(E_1)^k, \ldots, P(E_p)^k$. The more k will be important, the more the extensions with small weights will be minimized.
- The sum of the weight functions must be equal to one: we divide with the sum of the extensions weights
$$\frac{P(E_j)^k}{\sum_{i=1}^{p} P(E_i)^k} \tag{21}$$

This correction gives more importance to the extensions with high weight function, and less importance to the others. For the moment, we don't fix the value of the power k, we want to test different values.

Let us apply this correction on the example 2. We take the power function $f(x) = x^2$. The weight function 0.1 becomes $0.1^2 = 0.01$ and the weight function 0.5 becomes $0.5^2 = 0.25$.

We want the sum of the weight functions equal to 1. The sum of the weight functions with correction is

$$\sum_{i=1}^{6} P(E_i) = 0.3.$$

We obtain:

- Five extensions with the weight function $\frac{0.1^2}{0.3} = 0.04$.
- One extension with the weight function $\frac{0.5^2}{0.3} = 0.8$.

With the correction, we give more chances to the extension with the more important weight to be chosen.

6.3.3 Filtering of the Extensions with Small Weight Functions

To be sure to choose the most interesting extension and let the others aside (principle 1), we eliminate the extension with very small weight functions. We fix a threshold: the extensions with a weight function smaller than this threshold are removed.

6.4 Respect for Minimal Change

To respect the minimal change (principle 4), we define a general rule of minimal change. In this purpose, we remember the submarine behavior (random research trajectory, collision avoidance,...) at the time t. We call this behavior $Behavior(t)$, and the condition to be in this behavior: $Prerequisite$, which corresponds to the default prerequisite for this behavior. The general rule of minimal change is a default:

$$D_{min\_ch} = \frac{Behavior(t) \wedge Prerequisite(t) : Behavior(t+1)}{Behavior(t+1)} \quad (22)$$

This rule means: " If the prerequisite of the previous behavior are always true at the time t, and if it is possible to stay in this behavior at time $t+1$, the submarine can stay in this behavior at time $t+1$".

The preference coefficients $C_1 \ldots C_n$ can not be too important, in order to allow new behaviors. This rule gives more chances to an action already beginning and allows also persistency (principle 4).

7 Interface with the Simulator Framework and Results

An interface has been realized between NoMROD and the DCNS simulator framework ATANOR. ATANOR sends to NoMROD the information about the submarine at which we apply the behavior rules (course, speed, submersion, position) and about the enemy submarine (detection, position, speed).

NoMROD compiles the behavior rules, selects an extension and sends back the instruction of course, speed and submersion to the simulator framework. On the figure 1, we have an example of a run of NoMROD, interfaced with the simulator ATANOR.

We call Submarine 1 the submarine to which we apply the behavior rules of NoMROD. And we note Submarine 2 the enemy submarine, whose behavior is defined by ATANOR. The goal for the enemy submarine is to cross the patrol area without being detected. This transit is modeled with straight sections, which course is selected to match an average course.

In this scenario, the submarine 1 makes random research trajectory, because it has no detection. When it detects the enemy, the officer makes the sequence of actions: collision avoidance, elaboration of the solution, bypassing the enemy and tracking. The trajectory evaluated at the beginning is far from the real trajectory of the submarine 2. The manœuver of elaboration of the solution allows to obtain a better estimation of this trajectory.

We obtain an efficient program. To simulate a scenario of 2h40, the calculus time used by ATANOR and NoMROD is 6 seconds and the NoMROD program only used 20 % of this calculus time.

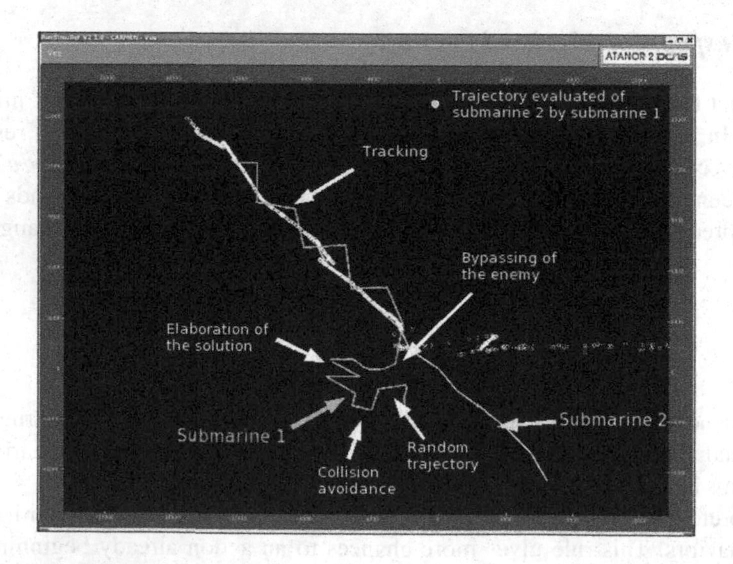

Fig. 1 Detection and tracking of a submarine

This scenario has been validated by submariners. They recognize the actions engaged when they detect a submarine. Moreover, we obtain a robust program. In the simulations, the trajectory evaluated by the submarine 1 about the submarine 2 give a bad estimation, far from the real trajectory of the submarine 2. In spite of this uncertain information, the modelling of the behavior allows to carry out the mission of tracking. Finally, this modelling was used to study the influence of the officer behavior on the operational performance of a submarine, with Monte-Carlo type statistical analyses.

8 Conclusion

The default logic allowed to formalize the behaviour rules of an officer in a submarine, by handling the incomplete, uncertain and revisable information on the environment.

With the use of Horn clauses and normal defaults, we obtain a simple, robust and efficient program, appropriate for military applications. Moreover, this is a work which can be applied generally, by using other clauses and other defaults.

We had to interface our work with the simulator framework ATANOR, in order to do statistical studies. We obtain good calculus time, so we can do such studies.

The obtained system compiles the available information and gives all the possibilities of actions with the extensions. To simulate the officer's choice, we defined a method to choose an extension:

- we defined weight functions for extensions, with preference coefficients on the defaults and a weighted sum;
- we defined a method to choose an extension thanks to these weight functions with a probabilistic technique: a random choice with corrections (power function and threshold) to be coherent with the principles which guide the officer's decisions.

We would like now to test another aggregation function: the Choquet integral, which allows to handle the interaction between the criteria. We must also work on a general method to attribute the value of the preference coefficients on the defaults and the character coefficients: we could use learning to attribute the best values.

Other methods exist to model the behavior: the behavior-based systems, introduced in robotics in 1980 [5], [11], [14], [1]. Another model has been tested: this model is based on the schema theory, developed by Arkin in 1989 [1]. This model, presented in [18], has some disadvantage:

- It is difficult to add a new behavior, because this method uses finite state automaton. If we want to add a new behavior, we have to review the finite state automaton.
- All the activated behavior are considered for the final instruction of the officer. We have a risk to obtain an incoherent behavior, which doesn't answer to the purpose of each activated behavior.

Our approach with the default logic allows to add easily a new behavior: we don't use Petri nets or states machines. The user will be able to add new rules, without having to care about rules previously established. So, we have the advantage to work with general rules, the defaults. We have just to define the mutual exclusions between the different behaviours. With the mutual exclusions, we have no risk to obtain incoherent behavior.

References

1. Arkin, R.: Motor schema-based mobile robot navigation. In: Proceedings of the IEEE International Conference on Robotics and Automation, pp. 264–271 (1987)
2. Ben Mena, S.: Introduction aux méthodes multicritères d'aide la décision. Biotchnol. Agron. Soc. Envion., 83–93 (2000)
3. Botto, J., Toulgoat, I., Audoly, C.: Modélisation de l'avantage tactique d'un sous-marin. In: 10ème Congrès Français d'Acoustique (2010)
4. Brewka, G.: Adding priorities and specificity to default logic. In: MacNish, C., Moniz Pereira, L., Pearce, D.J. (eds.) JELIA 1994. LNCS (LNAI), vol. 838, pp. 247–260. Springer, Heidelberg (1994)
5. Brooks, R.: A robust layered control system for a mobile robot. IEEE Journal of Robotics and Automation 2(1), 14–23 (1985)
6. Cordier, M., Siegel, P.: A temporal revision model for reasoning about world change. Journal of Intelligent Systems 9(1), 131–144 (1994)
7. Ferber: Les systèmes multi-agents. Vers une intelligence collective. InterEditions (1995)
8. Ginsberg, M., Smith, D.: Reasoning about action 1: a possible worlds approach. Readings in Non Monotonic Reasoning (1987)
9. Grabisch, M.: L'utilisation de l'intégrale de choquet en aide multicritère la décision. Newsletter of the European Working Group: Multicriteria Aid for Decisions (2006)
10. Grabisch, M., Labreuch, C.: Fuzzy measures and integrals in mcda. In: Figueira, J., Greco, S., Ehrgott, M. (eds.) Multiple Criteria Decision Analysis, pp. 563–608. Springer, Heidelberg (2005)
11. Maes, P.: The dynamics of action selection. In: Proceedings of the Eleventh International Joint Conference on Artificial Intelligence, vol. 2, pp. 991–997 (1989)
12. Nicolas, P., Garcia, L., Stephan, I.: Possibilistic stable models. In: International Joint Conferences on Artificial Intelligence, vol. 19, p. 248 (2005)
13. Petri, C.: Fundamentals of a theory of asynchronous information flow. In: 1st IFIP World Computer Congress, p. 386 (1962)
14. Pirjanian, P.: Behavior coordination mechanisms-state-of-the-art. Institute for Robotics and Intelligent Systems, School of Engineering, University of Southern California. Tech. Rep. IRIS-99-375 (1999)
15. Prouty, J.: Displaying uncertainty: a comparison between submarine subject matter experts. Ph.D. thesis, Naval postgraduate school, Monterey, California (2007)
16. Reiter, R.: A logic for default reasoning. Artificial intelligence 13(1–2), 81–132 (1980)
17. Sombé, L.: Raisonnement sur des informations incomplètes en intelligence artificielle, Teknea (1989)
18. Toulgoat, I., Botto, J., De lassus, Y., Audoly, C.: Modeling operator decision in underwater warfare performance simulations. In: Conference UDT, Cannes (2009)

19. Toulgoat, I., Siegel, P., Lacroix, Y., Botto, J.: Operator decision in naval action's simulations. In: NMR (2010)
20. Toulgoat, I., Siegel, P., Lacroix, Y., Botto, J.: Operator decision modeling in a submarine. In: COMPIT, p. 65 (2010)
21. Toulgoat, I., Siegel, P., Lacroix, Y., Botto, J.: Simulation du comportement d'un opérateur en situation de combat naval. JFPC (2010)
22. Vincke, P.: L'aide multicritère la décision. Ellipses (1989)
23. Winslett, M.: Reasoning about actions using a possible model approach. In: Proceedings of the 7th National Conference of AI, pp. 89–93 (1988)

Automatic Wrapper Adaptation by Tree Edit Distance Matching

Emilio Ferrara and Robert Baumgartner

Abstract. Information distributed through the Web keeps growing faster day by day, and for this reason, several techniques for extracting Web data have been suggested during last years. Often, extraction tasks are performed through so called wrappers, procedures extracting information from Web pages, e.g. implementing logic-based techniques. Many fields of application today require a strong degree of robustness of wrappers, in order not to compromise assets of information or reliability of data extracted.

Unfortunately, wrappers may fail in the task of extracting data from a Web page, if its structure changes, sometimes even slightly, thus requiring the exploiting of new techniques to be automatically held so as to adapt the wrapper to the new structure of the page, in case of failure. In this work we present a novel approach of *automatic wrapper adaptation* based on the measurement of similarity of trees through improved tree edit distance matching techniques.

1 Introduction

Web data extraction, during last years, captured attention both of academic research and enterprise world because of the huge, and still growing, amount of information distributed through the Web. Online documents are published in several formats but previous work primarily focused on the extraction of information from HTML Web pages.

Emilio Ferrara
University of Messina, Dept. of Mathematics, Via Ferdinando Stagno D'Alcontres, Salita Sperone, n. 31, Italy
e-mail: emilio.ferrara@unime.it

Robert Baumgartner
Lixto Software GmbH, Favoritenstrasse 16/DG, 1040 Vienna, Austria
e-mail: robert.baumgartner@lixto.com

I. Hatzilygeroudis and J. Prentzas (Eds.): Comb. of Intell. Methods and Appl., SIST 8, pp. 41–54.
springerlink.com

Most of the wrapper generation tools developed during last years provide to full support for users in building data extraction programs (a.k.a. wrappers) automatically and in a visual way. They can reproduce the navigation flow simulating the human behavior, providing support for technologies adopted to develop Web pages, and so on. Unfortunately, a problem still holds: wrappers, because of their intrinsic nature and the complexity of extraction tasks they perform, usually are strictly connected to the structure of Web pages (i.e. DOM tree) they handle. Sometimes, also slight changes to that structure can cause the failure of extraction tasks. A couple of wrapper generation systems try to natively avoid problems caused by minor changes, usually building more elastic wrappers (e.g. working with relative, instead of absolute, XPath queries to identify elements).

Regardless of the degree of flexibility of the wrapper generator, wrapper maintenance is still a required step of a wrapper life-cycle. Once the wrapper has been correctly developed, it could work for a long time without any malfunction. The main problem in the wrapper maintenance is that no one can predict when or what kind of changes could occur in Web pages.

Fortunately, local and minor changes in Web pages are much more frequent case than deep modifications (e.g. layout rebuilding, interfaces re-engineering, etc.). However, it could also be possible, after a minor modification on a page, that the wrapper keeps working but data extracted are incorrect; this is usually even worse, because it causes a lack of consistency of the whole data extracted. For this reason, state-of-the-art tools started to perform validation and cleansing on data extracted; they also provide caching services to keep copy of the last working version of Web pages involved in extraction tasks, so as to detect changes; finally, they notify to maintainers any change, letting possible to repair or rewrite the wrapper itself. Depending on the complexity of the wrapper, it could be more convenient to rewrite it from scratch instead of trying to find causes of errors and fix them.

Ideally, a robust and reliable wrapper should include directives to auto-repair itself in case of malfunction or failure in performing its task. Our solution of automatic wrapper adaptation relies on exploiting the possibility of comparing some structural information acquired from the old version of the Web page, with the new one, thus making it possible to re-induct automatically the wrapper, with a custom degree of accuracy.

The rest of the paper is organized as follows: in Section 2 we consider the related work on theoretical background and Web data extraction, in particular regarding algorithms, techniques and problems of wrapper maintenance and adaptation. Sections 3 covers the automatic wrapper adaptation idea we developed, detailing some interesting aspects of algorithms and providing some examples. Experimentation and results are discussed in Section 4. Section 5, finally, presents some conclusive considerations.

2 Related Work

Theoretical background on techniques and algorithms widely adopted in this work relies on several Computer Science and Applied Mathematics fields such as

Algorithms and Data Structures and Artificial Intelligence. In the setting of Web data extraction, especially algorithms on (DOM) trees play a predominant role. Approaches to analyze similarities between trees were developed starting from the well-known problem of finding the longest common subsequence(s) between two strings. Several algorithms were suggested, for example Hirshberg [4] provided the proof of correctness of three of them.

Soon, a strong interconnection between this problem and the similarity between trees has been pointed out: Tai [13] introduced the notion of *distance* as measure of the (dis)similarity between two trees and extended the notion of longest common subsequence(s) between strings to trees. Several *tree edit distance* algorithms were suggested, providing a way to transform a labeled tree in another one through local operations, like inserting, deleting and relabeling nodes. Bille [1] reported, in a comprehensive survey on the tree edit distance and related problems, summarizing approaches and analyzing algorithms.

Algorithms based on the tree edit distance usually are complex to be implemented and imply a high computational cost. They also provide more information than needed, if one just wants to get an estimate on the similarity. Considering these reasons, Selkow [12] developed a top-down trees isomorphism algorithm called *simple tree matching*, that establishes the degree of similarity between two trees, analyzing subtrees recursively. Yang [16] suggested an improvement of the simple tree matching algorithm, introducing weights.

During years, some improvements to tree edit distance techniques have been introduced: Shasha and Zhang [18] provided proof of correctness and implementation of some new parallelizable algorithms for computing edit distances between trees, lowering complexity of $O(|T_1| \cdot |T_2| \cdot \min(\mathit{depth}(T_1), \mathit{leaves}(T_1)) \cdot \min(\mathit{depth}(T_2), \mathit{leaves}(T_2)))$, for the non parallel implementation, to $O(|T_1| + |T_2|)$, for the parallel one; Klein [6], finally, suggested a fast method for computing the edit distance between unrooted ordered trees in $O(n^3 \log n)$. An overview of interesting applications of these algorithms in Computer Science can be found in Tekli et al. [14].

Literature on Web data extraction is manifold: Ferrara et al. [3] provided a comprehensive survey on application areas and used techniques, and Laender et al. [8] give a very good overview on wrapper generation techniques. Focusing on *wrapper adaptation*, Chidlovskii [2] presented some experimental results of combining and applying some grammatical and logic-based rules. Lerman et al. [9] developed a machine-learning based system for wrapper verification and reinduction in case of failure in extracting data from Web pages.

Meng et al. [10] suggested a new approach, called SG-WRAM (Schema-Guided WRApper Maintenance), for wrapper maintenance, considering that changes in Web pages always preserve syntactic features (i.e. data patterns, string lengths, etc.), hyperlinks and annotations (e.g. descriptive information representing the semantic meaning of a piece of information in its context).

Wong [15] developed a probabilistic framework to adapt a previously learned wrapper to unseen Web pages, including the possibility of discovering new attributes, not included in the first one, relying on the extraction knowledge related to the first wrapping task and on the collection of items gathered from the first Web page. Raposo et al. [11] already suggested the possibility of exploiting previously acquired information, e.g. queries results, to re-induct a new wrapper from an old one not working anymore, because of structural changes in Web pages.

Kim et al. [5] compared results of simple tree matching and a modified weighed version of the same algorithm, in extracting information from HTML Web pages; this approach shares similarities to the one followed here to perform adaptation of wrappers. Kowalkiewicz et al. [7] focused on robustness of wrappers exploiting absolute and relative XPath queries.

3 Wrapper Adaptation

As previously mentioned, our idea is to compare some helpful structural information stored by applying the wrapper on the original version of the Web page, searching for similarities in the new one.

3.1 Primary Goals

Regardless of the method of extraction implemented by the wrapping system (e.g. we can consider a simple XPath), elements identified and represented as subtrees of the DOM tree of the Web page, can be exploited to find similarities between two different versions.

In the simplest case, the XPath identifies just a single element on the Web page (Figure 1.A); our idea is to look for some elements, in the new Web page, sharing similarities with the original one, evaluating comparable features (e.g. subtrees, attributes, etc.); we call these elements *candidates*; among candidates, the one showing the higher degree of similarity, probably, represents the new version of the original element.

It is possible to extend the same approach in the common case in which the XPath identifies multiple similar elements on the original page (e.g. a XPath selecting results of a search in a retail online shop, represented as table rows, divs or list items) (Figure 1.B); it is possible to identify multiple elements sharing a similar structure in the new page, within a custom level of accuracy (e.g. establishing a threshold value of similarity). Section 4 discusses also these cases.

Once identified, elements in the new version of the Web page can be extracted as usual, for example just re-inducting the XPath. Our purpose is to define some rules to enable the wrapper to face the problem of automatically adapting itself to extract information from the new Web page.

We implemented this approach in a commercial tool [1]; the most efficient way to acquire some structural information about elements the original wrapper extracts, is to store them inside the definition of the wrapper itself. For example, generating *signatures* representing the DOM subtree of extracted elements from the original Web page, stored as a tree diagram, or a simple XML document (or, even, the HTML

[1] Lixto Suite, www.lixto.com

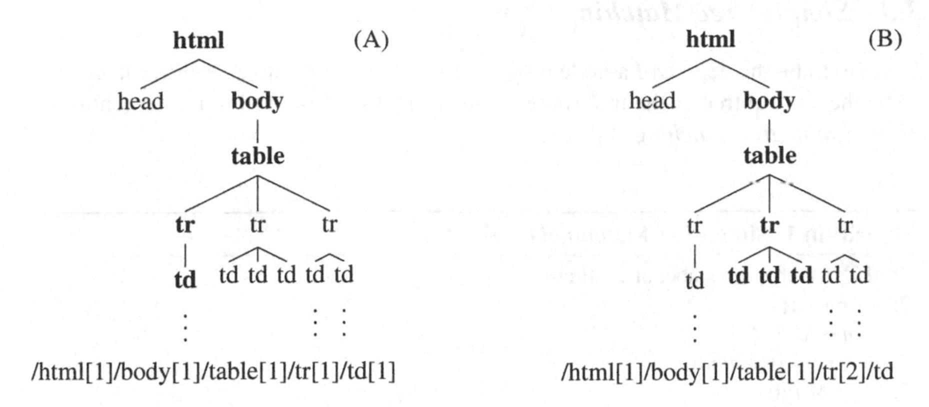

Fig. 1 Examples of XPaths over trees, selecting one (A) or multiple (B) items.

itself). This shrewdness avoids that we need to store the whole original page, ensuring better performances and efficiency.

This technique requires just a few settings during the definition of the wrapper step: the user enables the automatic wrapper adaptation feature and set an accuracy threshold. During the execution of the wrapper, if some XPath definition does not match a node, the wrapper adaptation algorithm automatically starts and tries to find the new version of the missing node.

3.2 Details

First of all, to establish a measure of similarity between two trees we need to find some comparable properties between them. In HTML Web pages, each node of the DOM tree represents an HTML element defined by a tag (or, otherwise, free text). The simplest way to evaluate similarity between two elements is to compare their *tag name*. Elements own some particular common attributes (e.g. *id*, *class*, etc.) and some type-related attributes (e.g. *href* for anchors, *src* for images, etc.); it is possible to exploit this information for additional checks and comparisons.

The algorithm selects candidates between subtrees sharing the same root element, or, in some cases, *comparable* -but not identical- elements, analyzing tags. This is very effective in those cases of deep modification of the structure of an object (e.g. conversion of tables in divs).

As discussed in Section 2, several approaches have been developed to analyze similarities between HTML trees; for our purpose we improved a version of the *simple tree matching* algorithm, originally led by Selkow [12]; we call it *clustered tree matching*. There are two important novel aspects we are introducing in facing the problem of the automatic wrapper adaptation: first of all, exploiting previously acquired information through a smart and focused usage of the tree similarity comparison; thus adopting a consolidated approach in a new field of application. Moreover, we contributed applying some particular and useful changes to the algorithm itself, improving its behavior in the HTML trees similarity measurement.

3.3 Simple Tree Matching

Let *d(n)* to be the degree of a node *n* (i.e. the number of first-level children); let T(i) to be the i-*th* subtree of the tree rooted at node T; this is a possible implementation of the *simple tree matching* algorithm:

Algorithm 1. SimpleTreeMatching(T', T'')

1: **if** T' has the same label of T'' **then**
2: $m \leftarrow d(T')$
3: $n \leftarrow d(T'')$
4: **for** $i = 0$ to m **do**
5: $M[i][0] \leftarrow 0$;
6: **for** $j = 0$ to n **do**
7: $M[0][j] \leftarrow 0$;
8: **for all** i such that $1 \leq i \leq m$ **do**
9: **for all** j such that $1 \leq j \leq n$ **do**
10: $M[i][j] \leftarrow \text{Max}(M[i][j-1], M[i-1][j], M[i-1][j-1] + W[i][j])$ where $W[i][j]$ = SimpleTreeMatching($T'(i-1)$, $T''(j-1)$)
11: return M[m][n]+1
12: **else**
13: return 0

Advantages of adopting this algorithm, which has been shown quite effective for Web data extraction [5, 17], are multiple; for example, the *simple tree matching* algorithm evaluates similarity between two trees by producing the maximum matching through dynamic programming, without computing inserting, relabeling and deleting operations; moreover, tree edit distance algorithms relies on complex implementations to achieve good performance, instead *simple tree matching*, or similar algorithms are very simple to implement.

The computational cost is $O(n^2 \cdot \max(leaves(T'), leaves(T'')) \cdot \max(depth(T'), depth(T'')))$, thus ensuring good performances, applied to HTML trees. There are some limitations; most of them are irrelevant but there is an important one: this approach can not match permutations of nodes. Despite this intrinsic limit, this technique appears to fit very well to our purpose of measuring HTML trees similarity.

3.4 Clustered Tree Matching

Let $t(n)$ to be the number of total siblings of a node n including itself:

Algorithm 2. ClusteredTreeMatching(T', T'')

```
1: {Change line 11 with the following code}
2: if m > 0 AND n > 0 then
3:     return M[m][n] * 1 / Max(t(T'), t(T''))
4: else
5:     return M[m][n] + 1 / Max(t(T'), t(T''))
```

In order to better reflect a good measure of similarity between HTML trees, we applied some focused changes to the way of assignment of a value for each matching node.

In the *simple tree matching* algorithm the assigned matching value is always 1. After leading some analysis and considerations on structure of HTML pages, our intuition was to assign a weighed value, with the purpose of attributing less importance to slight changes, in the structure of the tree, when they occur in deep sublevels (e.g. missing/added leaves, small truncated/added branches, etc.) and also when they occur in sublevels with many nodes, because these mainly represent HTML list of items, table rows, etc., usually more likely to modifications.

In the *clustered tree matching*, the weighed value assigned to a match between two nodes is 1, divided by the greater number of siblings with respect to the two compared nodes, considering nodes themselves (e.g. Figure 2.A, 2.B); thus reducing the impact of missing/added nodes.

Before assigning a weight, the algorithm checks if it is comparing two leaves or a leaf with a node which has children (or two nodes which both have children). The final contribution of a sublevel of leaves is the sum of assigned weighted values to each leaf (cfr. Code Line (4,5)); thus, the contribution of the parent node of those leaves is equal to its weighed value multiplied by the sum of contributions of its children (cfr. Code Line (2,3)). This choice produces an effect of *clustering* the process of matching, subtree by subtree; this implies that, for each sublevel of leaves the maximum sum of assigned values is 1; thus, for each parent node of that sublevel the maximum value of the multiplication of its contribution with the sum of contributions of its children, is 1; each cluster, singly considered, contributes with a maximum value of 1. In the last recursion of this top-down algorithm, the two roots will be evaluated. The resulting value at the end of the process is the measure of similarity between the two trees, expressed in the interval [0,1]. The closer the final value is to 1, the more the two trees are similar.

Let us analyze the behavior of the algorithm with an example, already used by [16] and [17] to explain the simple tree matching (Figure 2): 2.A and 2.B are two very simple generic rooted labeled trees (i.e. the same structure of HTML trees). They show several similarities except for some missing nodes/branches.

Applying the *clustered tree matching* algorithm, in the first step (Figure 2.A, 2.B) contributions assigned to leaves, with respect to matches between the two trees, reflect the past considerations (e.g. a value of $\frac{1}{3}$ is established for nodes (h), (i) and (j), although two of them are missing in 2.B). Going up to parents, the summation

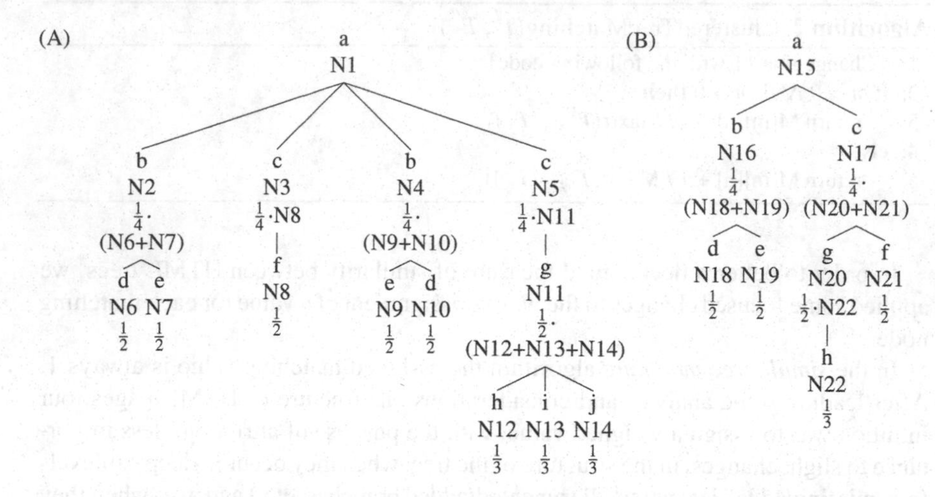

Fig. 2 *A* and *B* are two similar labeled rooted trees.

of contributions of matching leaves is multiplied by the relative value of each node (e.g. in the first sublevel, the contribution of each node is $\frac{1}{4}$ because of the four first-sublevel nodes in 2.A).

Once completed these operations for all nodes of the sublevel, values are added and the final measure of similarity for the two trees is obtained. Intuitively, in more complex and deeper trees, this process is iteratively executed for all the sublevels. The deeper a mismatch is found, the less its missing contribution will affect the final measure of similarity. Analogous considerations hold for missing/added nodes and branches, sublevels with many nodes, etc. Table 1 shows M and W matrices containing contributions and weights.

Table 1 *W* and *M* matrices for each matching subtree.

W	N18	N19
N6	$\frac{1}{2}$	0
N7	0	$\frac{1}{2}$

M	0	N18	N18-19
0	0	0	0
N6	0	$\frac{1}{2}$	$\frac{1}{2}$
N6-7	0	$\frac{1}{2}$	1

W	N18	N19
N9	0	$\frac{1}{2}$
N10	$\frac{1}{2}$	0

M	0	N18	N18-19
0	0	0	0
N9	0	0	$\frac{1}{2}$
N9-10	0	$\frac{1}{2}$	$\frac{1}{2}$

W	N8
N20	0
N21	$\frac{1}{2}$

M	0	N8
0	0	0
N20	0	0
N20-21	0	$\frac{1}{2}$

W	N12	N13	N14
N22	$\frac{1}{3}$	0	0

M	0	N12	N12-13	N12-14
0	0	0	0	0
N22	0	$\frac{1}{3}$	$\frac{1}{3}$	$\frac{1}{3}$

W	N11
N20	$\frac{1}{6}$
N21	0

M	0	N11
0	0	0
N20	0	$\frac{1}{6}$
N20-21	0	$\frac{1}{6}$

W	N2	N3	N4	N5
N16	$\frac{1}{4}$	0	$\frac{1}{8}$	0
N17	0	$\frac{1}{8}$	0	$\frac{1}{24}$

M	0	N2	N2-3	N2-4	N2-5
0	0	0	0	0	0
N16	0	$\frac{1}{4}$	$\frac{1}{4}$	$\frac{1}{4}$	$\frac{1}{4}$
N16-17	0	$\frac{1}{4}$	$\frac{3}{8}$	$\frac{3}{8}$	$\frac{3}{8}$

In this example, ClusteredTreeMatching(2.A, 2.B) returns a measure of similarity of $\frac{3}{8}$ (0.375) whereas SimpleTreeMatching(2.A, 2.B) would return a mapping value of 7; the main difference on results provided by these two algorithms is the following: our *clustered tree matching* intrinsically produces an absolute measure of similarity between the two compared trees; the *simple tree matching*, instead, returns the mapping value and then it needs subsequent operations to establish the measure of similarity.

Hypothetically, in the *simple tree matching* case, we could suppose to establish a good estimation of similarity dividing the mapping value by the total number of nodes of the tree with more nodes; indeed, a value calculated in this way would be linear with respect to the number of nodes, thus ignoring important information as the position of mismatches, the number of mismatches with respect to the total number of subnodes/leaves in a particular sublevel, etc.

In this case, for example, the measure of similarity between 2.A and 2.B, applying this approach, would be $\frac{7}{14}$ (0.5). A greater value of similarity could suggest, wrongly, that this approach is more accurate. Experimentation showed us that, the closer the measure of similarity is to reflect changes in complex structures, the higher the accuracy of the matching process is. This fits particularly well for HTML trees, which often show very rich and articulated structures.

The main advantage of using the *clustered tree matching* algorithm is that, the more the structure of considered trees is complex and similar, the more the measure of similarity will be accurate. On the other hand, for simple and quite different trees the accuracy of this approach is lower than the one ensured by the *simple tree matching*. But, as already underlined, the most of changes in Web pages are usually minor changes, thus *clustered tree matching* appears to be a valid technique to achieve a reliable process of automatic wrapper adaptation.

4 Experimentation

In this section we discuss some experimentation performed on common fields of application [3] and following results. We tried to automatically adapt wrappers, previously built to extract information from particular Web pages, after some -often minor- structural changes.

All the followings are real use cases: we did not modify any Web page, original owners did; thus re-publishing pages with changes and altering the behavior of old wrappers. Our will to handle real use cases limits the number of examples of this study. These real use cases confirmed our expectations and simulations on ad hoc examples we prepared to test the algorithms.

We obtained an acceptable degree of precision using the *simple tree matching* and a great rate of precision/recall using the *clustered tree matching*. Precision, Recall and F-Measure will summarize these results showed in Table 2. We focused on following areas, widely interested by Web data extraction:

- News and Information: Google News [2] is a valid use case for wrapper adaptation; templates change frequently and sometimes is not possible to identify elements with old wrappers.
- Web Search: Google Search [3] completely rebuilt the results page layout in the same period we started our experimentation [4]; we exploited the possibility of automatically adapting wrappers built on the old version of the *result page*.
- Social Networks: another great example of continuous restyling is represented by the most common social network, Facebook [5]; we successfully adapted wrappers extracting friend lists also exploiting additional checks performed on attributes.
- Social Bookmarking: building *folksonomies* and *tagging* contents is a common behavior of Web 2.0 users. Several Websites provide platforms to aggregate and classify sources of information and these could be extracted, so, as usual, wrapper adaptation is needed to face chages. We choose Delicious [6] for our experimentation obtaining stunning results.
- Retail: these Websites are common fields of application of data extraction and Ebay [7] is a nice real use case for wrapper adaptation, continuously showing, often almost invisible, structural changes which require wrappers to be adapted to continue working correctly.
- Comparison Shopping: related to the previous category, many Websites provide tools to compare prices and features of products. Often, it is interesting to extract this information and sometimes this task requires adaptation of wrappers to structural changes of Web pages. Kelkoo [8] provided us a good use case to test our approach.
- Journals and Communities: Web data extraction tasks can also be performed on the millions of online Web journals, blogs and forums, based on open source blog publishing applications (e.g. Wordpress [9], Serendipity [10], etc.), CMS (e.g. Joomla [11], Drupal [12], etc.) and community management systems (e.g. phpBB [13], SMF [14], etc.). These platforms allow changing templates and often this implies wrappers must be adapted. We lead the automatic adaptation process on Techcrunch [15], a tech journal built on Wordpress.

[2] http://news.google.com
[3] http://www.google.com
[4] http://googleblog.blogspot.com/2010/05/spring-metamorphosis-googles-new-look.html
[5] http://www.facebook.com
[6] http://www.delicious.com
[7] http://www.ebay.com
[8] http://shopping.kelkoo.co.uk
[9] http://wordpress.org
[10] http://www.s9y.org
[11] http://www.joomla.org
[12] http://drupal.org
[13] http://www.phpbb.com
[14] http://www.simplemachines.org
[15] http://www.techcrunch.com

Table 2 Experimental results.

		Simple Tree Matching			Clustered Tree Matching		
		Precision/Recall			Precision/Recall		
URL	threshold	true pos.	false pos.	false neg.	true pos.	false pos.	false neg.
news.google.com	90%	604	-	52	644	-	12
google.com	80%	100	-	60	136	-	24
facebook.com	65%	240	72	-	240	12	-
delicious.com	40%	100	4	-	100	-	-
ebay.com	85%	200	12	-	196	-	4
kelkoo.co.uk	40%	60	4	-	58	-	2
techcrunch.com	85%	52	-	28	80	-	-
Total	-	1356	92	140	1454	12	42
Recall	-	90.64%			97.19%		
Precision	-	93.65%			99.18%		
F-Measure	-	92.13%			98.18%		

We adapted wrappers for these 7 use cases considering 70 Web pages; Table 2 summarizes results obtained comparing the two algorithms applied on the same page, with the same configuration (threshold, additional checks, etc.). *Threshold* represents the value of similarity required to match two trees. The columns *true pos.*, *false pos.* and *false neg.* represent true and false positive and false negative items extracted from Web pages through adapted wrappers.

In some situations of deep changes (Facebook, Kelkoo, Delicious) we had to lower the threshold in order to correctly match the most of the results. Both the algorithms show a great elasticity and it is possible to adapt wrappers with a high degree of reliability; the *simple tree matching* approach shows a weaker recall value, whereas performances of the *clustered tree matching* are stunning (F-Measure greater than 98% is an impressive result). Sometimes, additional checks on nodes attributes are performed to refine results of both the two algorithms. For example, we can additionally include attributes as part of the node label (e.g. *id, name* and *class*) to refine results. Also without including these additional checks, the most of the time the false positive results are very limited in number (cfr. the Facebook use case).

Figure 3 shows a screenshot of the developed tool, performing an automatic wrapper adaptation task: in this example we adapted the wrapper defined for extracting Google news, whereas the original XPath was not working because of some structural changes in the layout of news. Elements identified by the original XPath are highlighted in red in the upper browser, elements highlighted in the bottom browser represent the recognized ones through the wrapper adaptation process.

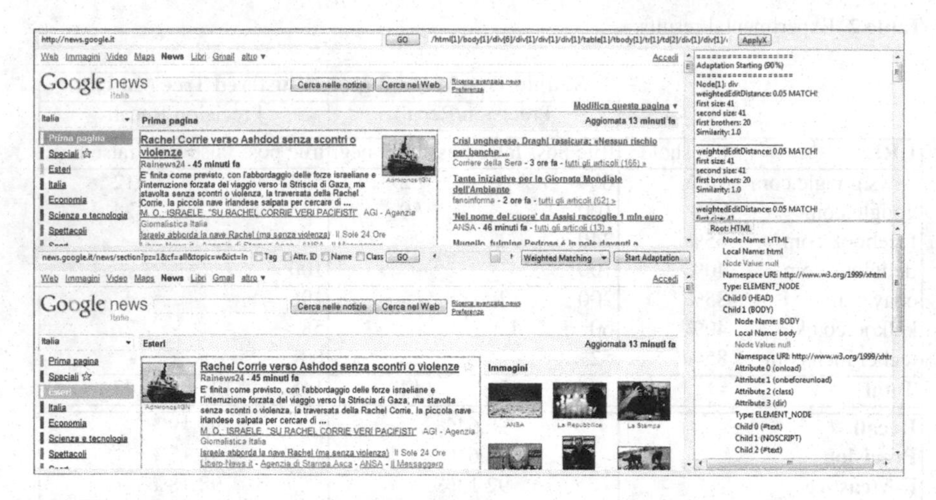

Fig. 3 An example of Wrapper Adaptation.

5 Conclusion

This work presents new scenarios, analyzing Wrapper Adaptation related problems from a novel point of view, introducing improvements to algorithms and new fields of application.

There are several possible improvements to our approach we can already imagine. First of all, it could be very interesting to extend the matching criteria we used, making the tree matching algorithm smarter. Actually, we already included features like analyzing attributes (e.g. *id*, *name* and *class*) instead of just comparing labels/tags or node types. The accuracy of the matching process benefits of these additional checks and it is possible, for example, to improve this aspect with a more complex matching technique, containing full path information, all attributes, etc.

It could be interesting to compare these algorithms, with other tree edit distance approaches working with permutations; although, intuitively, *simple tree matching* based algorithms can not handle permutations on nodes, maybe it is possible to develop some enhanced version which solves this limitation. Furthermore, just considering the tree structure can be limiting in some particular situations: if a new node has only empty textual fields (or, equally, if a deleted node had only empty fields) we could suppose its weight should be null. In some particular situation this inference works well, in some others, instead, it could provoke mismatches. It could also be interesting to exploit textual properties, nevertheless, not necessarily adopting Natural Language Processing techniques (e.g. using logic-based approaches, like regular expressions, or string edit distance algorithms, or just the length of strings – treating two nodes as equal only if the textual content is similar or of similar length).

The tree grammar could also be used in a machine learning approach, for example creating some tree templates to match similar structures or tree/cluster diagrams to classify and identify several different topologies of common substructures in Web pages. This process of simplification is already used to store a light-weight snapshot of elements identified by a wrapper applied on a Web page, at the time of extraction; actually, this feature allows the algorithm to work also without the original version of the page, but just exploiting some information about extracted items. This possibility opens new scenarios for future work on Wrapper Adaptation.

Concluding, the *clustered tree matching* algorithm we described is very extensible and resilient, so as ensuring its use in several different fields, for example it perfectly fits in identifying similar elements belonging to a same structure but showing some small differences among them. Experimentation on wrapper adaptation has already been performed inside a productive tool, the Lixto Suite, this because our approach has been shown to be solid enough to be implemented in real systems, ensuring great reliability and, generically, stunning results.

References

1. Bille, P.: A survey on tree edit distance and related problems. Theoretical Computer Science 337(1-3), 217–239 (2005), doi:10.1016/j.tcs.2004.12.030
2. Chidlovskii, B.: Automatic repairing of web wrappers. In: Proceedings of the 3rd international workshop on Web information and data management, p. 30. ACM Press, New York (2001)
3. Ferrara, E., Fiumara, G., Baumgartner, R.: Web Data Extraction, Applications and Techniques: A Survey. Technical Report (2010)
4. Hirschberg, D.S.: A linear space algorithm for computing maximal common subsequences. Communications of the ACM 18(6), 343 (1975)
5. Kim, Y., Park, J., Kim, T., Choi, J.: Web Information Extraction by HTML Tree Edit Distance Matching. In: Proceedings of the 2007 International Conference on Convergence Information Technology, vol. 1, pp. 2455–2460. IEEE, Los Alamitos (2007), doi:10.1109/ICCIT.2007.19
6. Klein, P.: Computing the edit-distance between unrooted ordered trees. In: Algorithms –ESA. LNCS, vol. 1461, pp. 1–1. Springer, Heidelberg (1998)
7. Kowalkiewicz, M., Kaczmarek, T., Abramowicz, W.: MyPortal: robust extraction and aggregation of web content. In: Proceedings of the 32nd International Conference on Very Large Data Bases, pp. 1219–1222 (2006)
8. Laender, A.H.F., Ribeiro-Neto, B.A., Da, A.S., Silva, J.S.: A brief survey of web data extraction tools. ACM Sigmod 31(2), 84–93 (2002), doi:10.1145/565117.565137
9. Lerman, K., Minton, S., Knoblock, C.: Wrapper maintenance: A machine learning approach. Journal of Artificial Intelligence Research 18, 149–181 (2003)
10. Meng, X., Hu, D., Li, C.: Schema-guided wrapper maintenance for web-data extraction. In: Proceedings of the 5th ACM international workshop on Web information and data management, pp. 1–8. ACM Press, New York (2003), doi:10.1145/956699.956701
11. Raposo, J., Pan, A., Álvarez, M., Viña, A.: Automatic wrapper maintenance for semi-structured web sources using results from previous queries. In: Proceedings of the 2005 ACM symposium on Applied computing - SAC 2005 , pp. 654–659. ACM Press, New York (2005), doi:10.1145/1066677.1066826
12. Selkow, S.: The tree-to-tree editing problem. Information Processing Letters 6(6), 184–186 (1977), doi:10.1016/0020-0190(77)90064-3
13. Tai, K.: The tree-to-tree correction problem. Journal of the ACM (JACM) 26(3), 433 (1979)

14. Tekli, J., Chbeir, R., Yetongnon, K.: An overview on XML similarity: Background, current trends and future directions. Computer Science Review 3(3), 151–173 (2009), doi:10.1016/j.cosrev.2009.03.001
15. Wong, T.: A Probabilistic Approach for Adapting Information Extraction Wrappers and Discovering New Attributes. In: Proceedings of the Fourth IEEE International Conference on Data Mining, pp. 257–264. IEEE, Los Alamitos (2004), doi:10.1109/ICDM.2004.10111
16. Yang, W.: Identifying syntactic differences between two programs. Software - Practice and Experience 21(7), 739–755 (1991)
17. Zhai, Y., Liu, B.: Web data extraction based on partial tree alignment. In: Proceedings of the 14th International Conference on World Wide Web, pp. 76–85. ACM Press, New York (2005), doi:10.1145/1060745.1060761
18. Zhang, K., Shasha, D.: Simple fast algorithms for the editing distance between trees and related problems. SIAM J. Comput. 18(6), 1245–1262 (1989)

Representing Temporal Knowledge in the Semantic Web: The Extended 4D Fluents Approach

Sotiris Batsakis and Euripides G.M. Petrakis

Abstract. Representing information that evolves in time in ontologies, as well as reasoning over static and dynamic ontologies are the areas of interest in this work. Building upon well established standards of the semantic Web and the 4D-fluents approach for representing the evolution of temporal information in ontologies, this work demonstrates how qualitative temporal relations that are common in natural language expressions (i.e., relations between time intervals like "before", "after", etc.) are represented in ontologies. Existing approaches allow for representations of temporal information, but do not support representation of qualitative relations and reasoning.

1 Introduction

Ontologies offer the means for representing high level concepts, their properties and their interrelationships. Dynamic ontologies will in addition enable representation of information evolving in time. In particular, dynamic ontologies are not only suitable for describing static scenes with static objects (e.g., objects in photographs) but also enable representation of events with objects and properties changing in time (e.g., moving objects in a video). Representation of both static and dynamic information in ontologies, as well as reasoning over static and dynamic ontologies are exactly the problems this work is dealing with.

Representation of dynamic features calls for mechanisms allowing representation of the notion of time (and of properties varying in time) [1]. Methods for achieving this goal include (among others), temporal description logics [11], temporal RDF [13], versioning [6], named graphs [18], reification, N-ary relations [2] and the 4D-fluent (perdurantist) approach [9] with the last being the most efficient. All

Sotiris Batsakis · Euripides G.M. Petrakis
Department of Electronic and Computer Engineering
Technical University of Crete (TUC) Chania, Crete, GR-73100, Greece
e-mail: batsakis@softnet.tuc.gr, petrakis@intelligence.tuc.gr

I. Hatzilygeroudis and J. Prentzas (Eds.): Comb. of Intell. Methods and Appl., SIST 8, pp. 55–69.
springerlink.com

approaches suffer from data redundancy as several objects are created for each binary relationship changing in time (i.e., for each new event, a new temporal object and an additional binary relationship for each temporal property of this object is created and associated with existing classes) thus complicating the ontology. Also, adding a time argument to binary relationships may (as in reification and named graphs) complicate application of OWL language constructs (e.g., cardinality constraints, inverse, transitive relations are no longer applicable) thus limiting OWL expressivity and obstructing reasoning. The 4D fluents approach, still suffers from data redundancy but maintains OWL expressiveness and reasoning support (i.e., an OWL reasoner such as Pellet can still be applied to fully exploit OWL semantics over the 4D fluent representation). However, time and temporal constructs representing the evolution of binary relationships in time, still offer additional semantics which can be exploited by applying additional rules (e.g., rules on Allen relationships). This is also a problem this work is dealing with.

Reasoning on temporal knowledge is still an active research area and has been investigated previously in other domains (temporal logics [11], temporal data bases [10]). To the best of our knowledge this is the first work to address this problem within the context of ontologies. More specifically, we show how results from previous research efforts [17, 28, 25] can be ported into ontological representations such as the extended 4D fluents representation proposed in this work.

In our earlier work [4] we showed how temporal information (also the evolution of temporal concepts) can be represented effectively in OWL. Concepts varying in time are represented as 4-D dimensional objects, with the 4-th dimension being the time. This work extends this approach in certain ways: The 4-D fluents mechanism is enhanced with qualitative (in addition to quantitative) temporal expressions allowing for the representation of temporal intervals with unknown starting and ending points by means of their relation (e.g., "before", "after") to other time intervals. Adding reasoning support to the above representation is also a contribution of the present work: A set of inference rules is proposed whose purpose is to assert additional implied facts into the knowledge base (i.e., determine the temporal relation between two events given their relations with a third one). Reasoning becomes feasible by using a tractable subset of the set of Allen's relationships [17]. Specifically, the reasoning mechanism incorporates rules for inferring certain temporal relations from existing ones using additional axioms based on compositions of Allen relations and by checking temporal assertions for consistency (i.e., path consistency checking is implemented).

Adding query support to the extended 4D fluent representation is an additional contribution of this work. More specifically, we extend the TOQL query language [4] to handle qualitative temporal relationships and the extended 4D fluent representation.

Related work in the field of knowledge representation is discussed in Section 2. This includes issues related to representing and reasoning over information evolving in time. The temporal representation model is presented in Section 3 and the corresponding reasoning mechanism in Section 3.1, followed by evaluation in Section 4 and conclusions and issues for future work in Section 5.

2 Background and Related Work

Several representation languages are defined for the Semantic Web, the most important of them are referred to as the OWL-family [7, 22] of ontology languages for ontology building and knowledge representation. Representation languages such as RDF, OWL (which is based on description logics), the same as frame-based and object-oriented languages (F-logic) are all based on binary relations. Binary relations simply connect two instances (e.g., an employee with a company) without any temporal information. Nevertheless, representation of time using OWL is feasible, although complicated [2, 9].

The OWL-Time temporal ontology [5] describes the temporal content of Web pages and the temporal properties of Web services. Apart from language constructs for the representation of time in ontologies, there is still a need for mechanisms for the representation of the evolution of concepts (e.g., events) in time. This is related to the problem of the representation of time in temporal (relational and object oriented) databases. Existing methods are relying mostly on temporal Entity Relation (ER) models [10] taking into account valid time (i.e., time interval during which a relation holds), transaction time (i.e., time at which a database entry is updated) or both. Also time is represented by time instants, intervals or finite sets of intervals. However, representation of time in OWL differs because (a) OWL semantics are not equivalent to ER model semantics (e.g., OWL adopts the *Open World Assumption* while ER model adopts the *Closed World Assumption*) and (b) relations in OWL are restricted to binary ones. Representation of time in the Semantic Web can be achieved using *Temporal Description logics (TDLs)* [11, 12], *Reification*, *N-ary relations* [2], *temporal RDF* [13], *Versioning* [6], *named graphs* [18] or *4D-fluents* [9].

Temporal Description Logics (TDLs) extend standard description logics (DLs) that form the basis for semantic Web standards with additional constructs such as "always in the past", "sometime in the future". TDLs offer additional expressive capabilities over non temporal DLs and retain decidability (with an appropriate selection of allowable constructs) but they require extending OWL syntax and semantics with additional temporal constructs. Representing information regarding specific time points requires support for concrete domains, resulting to the proliferation of objects [11].

Temporal RDF [13] proposes extending RDF by labeling properties with the time interval they hold. This approach also requires extending the syntax and semantics of the standard RDF, although representation over RDF (e.g., using reification) can be achieved. Note that Temporal-RDF cannot express incomplete information , by means of qualitative relations.

Reification is a general purpose technique for representing *n*-ary relations using a language such as OWL that permits only binary relations. Specifically, an *n*-ary relation is represented as a new object that has all the arguments of the *n*-ary relation as objects of properties. For example if the relation *R* holds between objects *A* and *B* at time *t*, this is expressed as *R(A,B,t)*. Furthermore, in OWL using reification this is expressed as a new object with *R*,*A*,*B* and *t* being objects of properties. Fig. 1 illustrates the relation *WorksFor(Employee, Company, TimeInterval)* representing

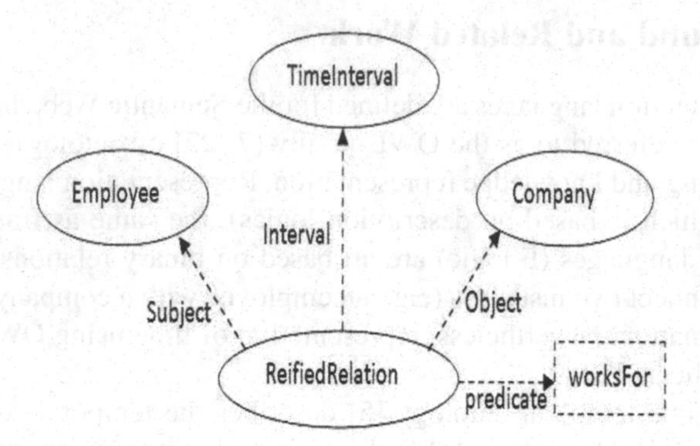

Fig. 1 Example of Reification

the fact that an employee works for a company during a time interval. Reification suffers mainly from two disadvantages: (a) data redundancy, because a new object is created whenever a temporal relation has to be represented (this problem is common to all approaches based on non temporal Description Logics such as OWL-DL) and (b) offers limited OWL reasoning capabilities [9] since relation R is represented as the object of a property thus OWL semantics over properties are no longer applicable (i.e., the properties of a relation are no longer associated directly with the relation itself).

N-ary relations is also a general purpose technique that represents an n-ary relation using an additional object. In contrast to reification, the n-ary relation is not represented as the object of a property but as two properties each related with the new object. These two objects are related to each other with an n-ary relation. This is also illustrated in Fig.2. This approach requires only one additional object for every temporal interval, maintains property semantics but suffers from data redundancy in the case of inverse and symmetric properties [2] (e.g., the inverse of a relation is added explicitly twice instead of once as in 4D-fluents).

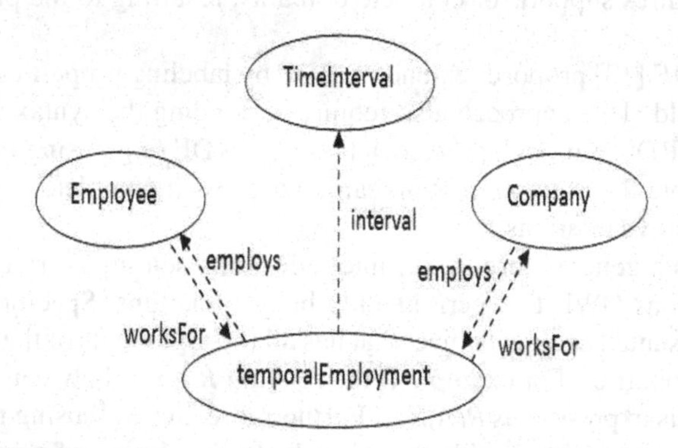

Fig. 2 Example of N-ary Relations

Versioning [6] suggests that the ontology has different versions (one per instance of time). When a change takes place, a new version is created. Versioning suffers from several disadvantages: (a) changes even on single attributes require that a new version of the ontology be created leading to information redundancy (b) searching for events occurred at time instances or during time intervals requires exhaustive searches in multiple versions of the ontology, (c) it is not clear how the relation between evolving classes is represented. Furthermore, ontology languages such as OWL [7] are based on binary relations (relations connecting two instances) with no time dimension regarding ontology versions.

Named Graphs [18] represent the temporal context of a property by inclusion of a triple representing the property in a named graph (i.e., a subgraph into the RDF graph of the ontology specified by a distinct name). The default (i.e., main) RDF graph contains definitions of interval start and end points for each named graph, thus a property is stored in a named graph with start and end points corresponding to the time interval that the property holds. Named graphs are not part of the OWL specification [24] (i.e., there are not OWL constructs translated into named graphs) and they are not supported by OWL reasoners.

The *4D-fluent* (perdurantist) approach [9] shows how temporal information and the evolution of temporal concepts can be represented effectively in OWL. Concepts in time are represented as 4-dimensional objects with the 4th dimension being the time. Time instances and time intervals are represented as instances of a *time interval* class which in turn is related with time concepts varying in time. Changes occur on the properties of the temporal part of the ontology keeping the entities of the static part unchanged. The 4D-fluent approach still suffers from data redundancy but in contrast to other approaches it maintain full OWL expressiveness and reasoning support. N-ary relations[2] is considered to be an alternative to the 4-D fluents approach, although the 4-D fluents representation where the property is holding among two timeslices of objects and not between the two objects and the intermediate object representing their relation may seems more natural to users. TOWL [23] is a temporal representation approach based on 4-D fluents that extends OWL syntax with temporal concepts and supports quantitative time intervals.

3 Extended 4D Fluents Approach

Following the approach by Welty and Fikes [9], to add time dimension to an ontology, classes *TimeSlice* and *TimeInterval* with properties *tsTimeSliceOf* and *tsTimeInterval* are introduced. Class *TimeSlice* is the domain class for entities representing temporal parts (i.e., "time slices") and class *TimeInterval* is the domain class of time intervals. A time interval holds the temporal information of a time slice. Property *tsTimeSliceOf* connects an instance of class *TimeSlice* with an entity, and property *tsTimeInterval* connects an instance of class *TimeSlice* with an instance of class *TimeInterval*. Properties having a time dimension are called fluent properties and connect instances of class *TimeSlice*.

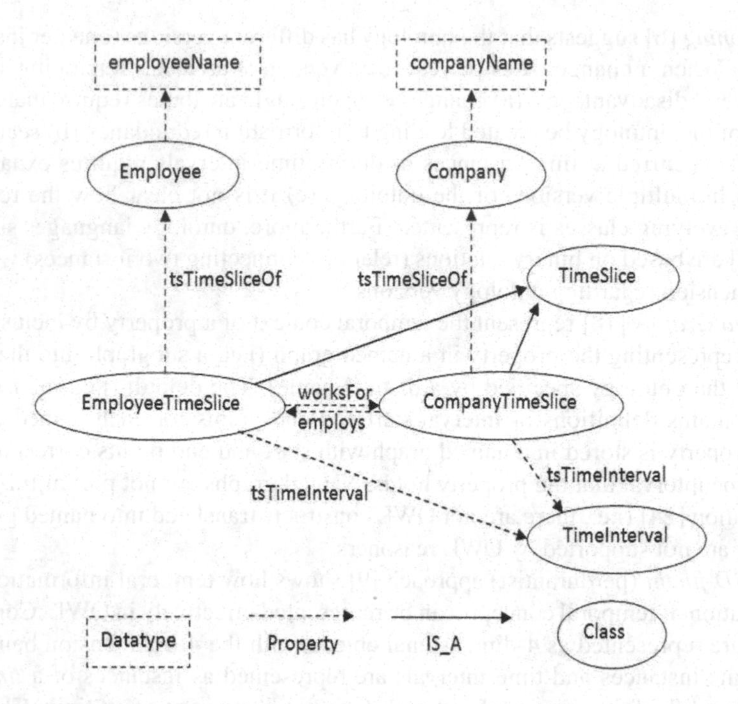

Fig. 3 Dynamic Enterprise Ontology

Fig. 3 illustrates a temporal ontology with classes *Company* with datatype property *companyName* and *Employee* with datatype property *employeeName*. In this example, *CompanyName* and *EmployeeName* are static properties (their value do not change in time), while properties *employs* and *worksFor* (i.e., inverse of employs) are dynamic (fluent) properties whose values may change in time. Because they are fluent properties, their domain (and range) is of class *TimeSlice*. *CompanyTimeSlice* and *EmployeeTimeslice* are instances of class *TimeSlice* and are provided to denote that the domain of properties *worksFor* and *employs*, are time slices restricted to be slices of a specific class. For example, the domain of property *employs* is not class *TimeSlice* but it is restricted to instances that are time slices of class *Company*.

The 4-D fluent mechanism forms the basis of the proposed temporal ontology representation. In this work, the 4D-fluent representation is enhanced with qualitative temporal relations holding between time intervals whose starting and ending points are not specified. This is implemented by introducing temporal relationships as object relations between time intervals. This can be one of the 13 pairwise disjoint Allen's relations [17] of Fig. 4.

By allowing for qualitative relations the expressive power of the representation increases. Temporal RDF and 4-D fluents both require closed temporal intervals for the representation of temporal information, while semiclosed and open intervals can't be represented effectively in a formal way. If their endpoints are unknown, ad-hoc approaches [18] that handle open intervals by extending their start or end

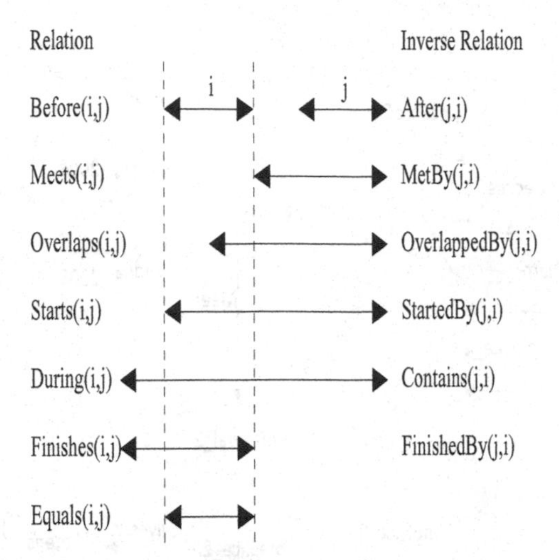

Fig. 4 Allen's Temporal Relations

point infinitely are not appropriate, since lack of knowledge (about their endpoints) is interpreted as if a property always holds in the past or future. In this work, this is handled by Allen relations: for example, if interval $t1$ is known and $t2$ is unknown but we know that $t2$ starts when $t1$ ends, then we can assert that $t2$ is *met by* $t1$. Likewise, if an interval $t3$ with unknown endpoints is introduced and $t3$ is *before* $t1$ then, using compositions of Allen relations [17], we infer that $t3$ is *before* $t2$ although both interval's endpoints are unknown and their relation is not represented explicitly in the ontology. Semiclosed intervals can be handled in a similar way. For example, if $t1$ starts at time point 1, still holds at time point 2, but it's endpoint is unknown, we assert that $t1$ is *started by* interval $t2$:$[1,2]$. Fig.5 illustrates the dynamic ontology schema representing the scenario "George lived in Crete from 2004 to 2010 and then he moved to Athens". In this example, we don't know whether George still lives in Athens.

Overall, the model demonstrates enhanced expressivity compared to previous approaches [18, 19, 23, 15] by combining 4D-fluents [9] with Allen's temporal relations, their formal semantics and composition rules as defined in [17].

3.1 Temporal Reasoning

Reasoning is realized by introducing a set of SWRL [27] rules operating on temporal intervals. Reasoners that support DL-safe rules such as Pellet [16] can be used for inference and consistency checking over temporal relations. In addition to reasoning applying on temporal relations, the Pellet reasoner is applied on the ontology schema to infer additional facts using OWL semantics (e.g., facts due symmetric relationships and class-subclass relationships).

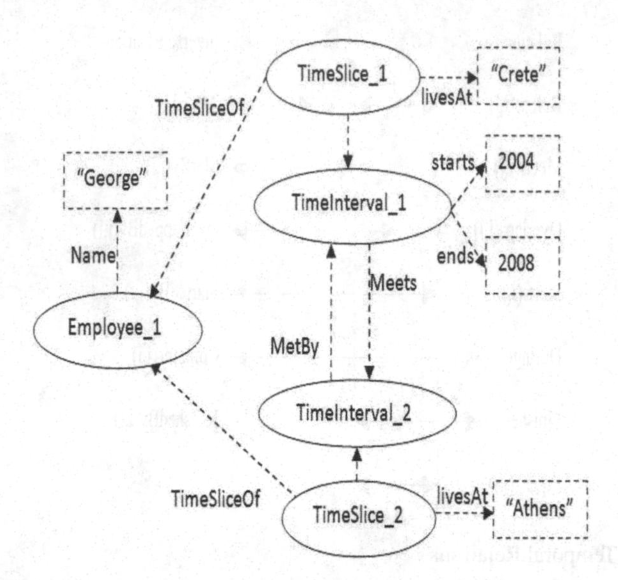

Fig. 5 Instantiation example

The temporal reasoning rules are based on composing pairs of basic Allen's relations of Fig. 4 as defined in [17]. The composition table of basic Allen's relations is presented in Table 1. Relations BEFORE, AFTER, MEETS, METBY, OVERLAPS, OVERLAPPEDBY, DURING, CONTAINS, STARTS, STARTEDBY, ENDS, ENDEDBY and EQUALS are represented using symbols B, A, M, Mi, O, Oi, D, Di, S, Si, F, Fi and = respecively. Compositions with EQUALS are not presented since these compositions keep the initial relations unchanged. The composition table represents the result of the composition of two Allen relations. For example, if relation $R1$ holds between $interval1$ and $interval2$ and relation $R2$ holds between $interval2$ and $interval3$ then the entry of the Table 1 corresponding to line $R1$ and column $R2$ denotes the possible relation(s) holding between $interval1$ and $interval3$. Not all compositions yield a unique relation as a result. For example the composition of relations $During$ and $Meets$ yields the relation $Before$ as result while the composition of relations *Overlaps* and *During* yields three possible relations *Starts, Overlaps and During*. Rules corresponding to compositions of relations *R1, R2* yielding unique relations $R3$ as a result can be represented using SWRL as follows:

$$R1(x,y) \land R2(y,z) \rightarrow R3(x,z)$$

An example of temporal inference rule is the following:

$$DURING(x,y) \land MEETS(y,z) \rightarrow BEFORE(x,z)$$

Rules yielding a set of possible relations as a result can't be represented in SWRL since disjunctions of atomic formulas are not permitted as a rule head. Instead,

Table 1 Composition Table for Allen's temporal relations.

	B	A	D	Di	O	Oi	M	Mi	S	Si	F	Fi
B	B	B,A,D,Di,O,Oi M,MiS,Si,F,Fi, Eq	B,O,M, D,S	B	B	B,O,M,D,S	B	B,O,M,D, S	B	B	B,O,M, D,S	B
A	B,A,D,Di, O,Oi,M, Mi,S,Si,F, Fi,Eq	A	A,Oi,Mi D,F	A	A,Oi, Mi,D,F	A	A,Oi,Mi, D,F	A	A,Oi,Mi D,F	A	A	A
D	B	A	D	B,A,D,Di,O,Oi M,MiS,Si,F,Fi, Eq	B,O,M, D,S	A,Oi,Mi,D,F	B	A	D	A,Oi,Mi D,F	D	B,O,M, D,S
Di	B,O,M,Di, Fi	A,Oi,Di,Mi,Si	O,Oi,D, Di,S,Si, F,Fi,Eq	Di	O,Di,Fi	Oi,Di,Si	O,Di,Fi	Oi,Di,Si	O,Di,Fi	Di	Oi,Di,Si	Di
O	B	A,Oi,Di,Mi,Si	O,D,S	B,O,M,Di,Fi	B,O,M	O,Oi,D,Di,S,Si, F,Fi,Eq	B	Oi,Di,Si	O	O,Di,Fi	O,D,S	B,O,M
Oi	B,O,M,Di, Fi	A	Oi,D,F	A,Oi,Di,Mi,Si	O,Oi,D, Di,S,Si, F,Fi,Eq	A,Oi,Mi	O,Di,Fi	A	Oi,D,F	Oi,A,Mi	Oi	Oi,Di,Si
M	B	A,Oi,Di,Mi,Si	O,D,S	B	B	O,D,S	B	F,Fi,Eq	M	M	O,D,S	B
Mi	B,O,M,Di, Fi	A	Oi,D,F	A	Oi,D,F	A	S,Si,Eq	A	Oi,D,F	A	Mi	Mi
S	B	A	D	B,O,M,Di,Fi	B,O,M	Oi,D,F	B	Mi	S	S,Si,Eq	D	B,O,M
Si	B,O,M,Di, Fi	A	Oi,D,F	Di	O,Di,Fi	Oi	O,Di,Fi	Mi	S,Si,Eq	Si	Oi	Di
F	B	A	D	A,Oi,Di,Mi,Si	O,D,S	A,Oi,Mi	M	A	D	A,Oi,Mi	F	F,Fi,Eq
Fi	B	A,Oi,Di,Mi,Si	O,D,S	Di	O	Oi,Di,Si	M	Oi,Di,Si	O	Di	F,Fi,Eq	Fi

disjunctions of relations are represented using new relations whose compositions must also be defined and asserted into the knowledge base. For example, if the relation *DOS* represents the disjunction of relations *During, Overlaps* and *Starts*, then the composition of *Overlaps* and *During* can be represented as follows:

$$OVERLAPS(x,y) \wedge DURING(y,z) \rightarrow DOS(x,z)$$

Note that the set of possible disjunctions over all basic Allen's relations is 2^{13} but subsets of this set that are closed under composition (i.e., compositions of relation pairs from this subset yield also a relation in this subset) do exist [25, 28]. In this work we use the tractable subset introduced in [28].

In addition to the above, the following axioms are also asserted into the knowledge base:

- Four transitivity axioms (for the relations BEFORE, FINISHEDBY, CONTAINS, STARTEDBY).

- Six inverse axioms (relations AFTER, METBY, OVERLAPPEDBY, STARTEDBY, CONTAINS and FINISHEDBY are the inverses of BEFORE, MEETS, OVERLAPS, STARTS, DURING and FINISHES respectively).
- One equality axiom (relation EQUALS).
- Rules defining the relation holding between two intervals with known starting and ending points (e.g., if ending of *interval1* is smaller than the start of *interval2* the *interval1* is *before interval2*) are part of the ontology as well.

Notice that, starting and ending points of intervals are represented using concrete datatypes such as *xsd:date* that support ordering relations. Axioms concerning relations that represent disjunctions of basic relations are defined using the corresponding axioms for these basic relations. Specifically, compositions of disjunctions of basic relations are defined as the disjunction of the compositions of these basic relations. For example the composition of relation *DOS* (representing the disjunction of *During, Overlaps and Starts*), and the relation *During* yields the relation *DOS* as a result as follows:

$$DOS \circ During \rightarrow (During \vee Overlaps \vee Starts) \circ During \rightarrow$$

$$(During \circ During) \vee (Overlaps \circ During) \vee (Starts \circ During)$$
$$\rightarrow (During) \vee (During \vee Overlaps \vee Starts) \vee (During)$$
$$\rightarrow During \vee Starts \vee Overlaps \rightarrow DOS$$

The symbol $\circ$ denotes composition of relations, and compositions of basic (non-disjunctive) relations are defined using Table 1. Similarly, the inverse of a disjunction of basic relations is the disjunction of the inverses of these basic relations as presented in Fig. 4. For example the inverse of the disjunction of relations $Before$ and $Meets$ is the disjunction of the inverse relations of $Before$ and $Meets$ ($After$ and $MetBy$ respectively).

By applying compositions of relations the implied relations may be inconsistent. Consistency checking is achieved using path consistency [14, 25, 28]. Path consistency is implemented by consecutive applications of the following formula:

$$\forall x,y,k\, R_s(x,y) \leftarrow R_i(x,y) \cap (R_j(x,k) \circ R_k(k,y))$$

representing intersection of compositions of relations with existing relations (the symbol $\cap$ denotes intersection and the symbol $\circ$ denotes composition and symbols R_i, R_j, R_k, R_s denote Allen relations). The formula is applied until a fixed point is reached (i.e., application of rules doesn't yield new inferences) or until the empty set is reached, implying that the ontology is inconsistent.

An additional set of rules defining the result of intersection of relations holding between two intervals are also introduced. These rules have the form:

$$R1(x,y) \wedge R2(x,y) \rightarrow R3(x,y)$$

where $R3$ can be the empty relation. For example the intersection of relation *DOS* (represents the disjunction of *During, Overlaps and Starts*), and the relation *During* yields the relation *During* as result:

$$DOS(x,y) \wedge During(x,y) \rightarrow During(x,y)$$

Intersection of relations *During* and *Starts* yields the empty relation, and an inconsistency is detected:

$$Starts(x,y) \wedge During(x,y) \rightarrow \bot$$

Notice that, using the full set of 2^{13} relations leads to intractability [29]. Tractable subsets of relations that polynomial time algorithms such as path-consistency are sound and complete (while these algorithms are approximation algorithms in the case of the full Allen algebra) do exist [25, 28, 30]. The largest such set (corresponding to the maximal tractable subset of Allen relations containing all basic relations when applying the path consistency method) comprises of 868 relations [25]. Tractable subsets of Allen relations containing 83 or 188 relations [28] can be used for reasoning as well, offering reduced expressivity but increased efficiency over the maximal subset of [25].

An ontology based on a set containing 83 relations (i.e., the *continuous endpoint subclass* presented in [28]) has been implemented in this work. Other relations corresponding to disjunctions of basic relations that are not supported (i.e., they don't belong to the subset referred to above) can't be asserted into the ontology. In [28] reasoning regarding time instants in addition to intervals is presented as well. Specifically qualitative relations regarding instants form a tractable set if the relation $\neq$ (i.e., a temporal instant is before *or* after another instant) is excluded. Reasoning regarding relations between interval and instants is achieved by translating interval relations to relations regarding their endpoints as specified in [17].

3.2 Querying Temporal Information

Querying temporal information over the semantic Web using general purpose languages such as [8] and SeRQL [3] is a tedious task. Recent work on query languages for temporal ontologies include TOQL [4] (extended with spatial operators at [33]) and t-SPARQL [18] using 4-D fluents and named graphs respectively for the representation of temporal information. Notice that, t-SPARQL suggests using named graphs as the underlying representation mechanism of temporal information and therefore, does not preserve OWL expressiveness, has no reasoning support and does not support representation of qualitative temporal expressions. TOQL handles all these issues. In this work TOQL is used for querying the temporal ontology.

TOQL is a query language that treats classes and properties of an ontology almost like tables and columns of a database. The language is enhanced with a set of temporal operators (i.e., the AT and Allen operators). TOQL follows an SQL-like syntax (SELECT-FROM-WHERE) and supports SQL operators and constructs such as LIMIT, OFFSET, AND, OR,MINUS, UNION, UNION ALL, INTERSECT, EXISTS, ALL, ANY, IN.

TOQL also introduces clause "AT" which compares a fluent property (i.e., the time interval in which the property is true) with a time period (time interval) or time point and returns fluents holding true at the specified time interval, thus enabling temporal queries without requiring familiarity with the underlying representation mechanism for the end user. For example the following TOQL query retrieves the name of the company employee "x" was working for, from time=3 to time=5:

> **SELECT** Company.companyName
> **FROM** Company, Employee
> **WHERE** Company.hasEmployee:Employee AT(3,5)
> AND Employee.employeeName LIKE "x"

The following Allen operators are also supported: BEFORE, AFTER, MEETS, METBY, OVERLAPS, OVERLAPPEDBY, DURING, CONTAINS, STARTS, STARTEDBY, ENDS, ENDEDBY and EQUALS, representing the corresponding relations holding between two time intervals specified either using quantitative (i.e., interval with specified end points) description or qualitative Allen relations. The following query retrieves the name of the company that hired employee "x" and *then* employee "y":

> **SELECT** Company.companyName
> **FROM** Company, Employee AS E1, Employee AS E2
> **WHERE** Company.hasEmployee:E1
> BEFORE Company.hasEmployee:E2
> AND E1.employeeName like "x"
> AND E1.employeeName LIKE "y"

In this work, extending TOQL to support queries over qualitative relations required certain modifications to the language. The basic SQL syntax remains the same, however, Allen operators aren't translated to comparisons of interval endpoints as in [4] but to Allen relations holding between intervals after reasoning is applied. The AT operator in [4] requires that interval endpoints are defined. Here, we introduce two additional operators namely *ALWAYS_AT* and *SOMETIME_AT* querying for fluents holding always during the interval in question and some time in the interval in question respecively. The AT operator in [4] corresponds to the proposed ALWAYS_AT operator. Specifically, the ALWAYS_AT operator returns fluents holding at intervals that EQUALS, CONTAINS, STARTEDBY or ENDEDBY the interval in question. The SOMETIME_AT operators returns fluents holding at intervals that OVERLAP, OVERLAPPEDBY, START, STARTEDBY, END, ENDEDBY, EQUAL, CONTAIN or DURING the interval in question. These semantics in conjunction with the reasoning mechanism will allow for application of the operators on qualitative intervals in addition to quantitative ones that are supported by the AT operator.

4 Evaluation

The resulting OWL ontology is characterized by *SHRIF(D)* DL expressivity and it is decidable since it doesn't contain role inclusion axioms with cyclic dependences [21] (role axioms in the ontology are restricted to disjointness, transitivity and inverse axioms). Adding the set of temporal qualitative rules of Sec. 3.1 retains decidability since rules are DL-safe rules as defined at [26, 31] and they apply only on named individuals of the ontology Abox using Pellet (which support DL-safe rules [32]). Furthermore, computing the rules has polynomial time complexity since a tractable subset of Allen's relations is used.

As shown in [14, 25, 28], by restricting the supported relations set to a tractable subset of Allen's algebra, path consistency has $O(n^5)$ time complexity (with n being the number of intervals). Also, any time interval can be related with every other interval by at most k relations, where k is the size of the set of supported relations. Therefore, for n intervals, using $O(k^2)$ rules, at most $O(kn^2)$ relations can be asserted into the knowledge base. Note that, extending the model for the full set of relations would result into an intractable reasoning procedure.

An alternative approach towards implementing a temporal reasoner would be to extend Pellet to handle a (tractable) relations set, along with the supported axioms and path consistency checking, similarly to the way PelletSpatial [20] implements reasoning over RCC-8 topologic relations. This approach has the following advantages: (a) The underlying representation is more simple since only the 13 Basic Allen relations have to be defined and (b) certain improvements regarding efficiency and scalability can be added. On the other hand, this approach requires additional software to handle the ontology, while our approach requires only standard semantic Web tools such as Pellet and SWRL. Because reasoning is part of the ontology model, maintenance of the ontology requires that changes are applied to the ontology only and not to the reasoner (other approaches such as [20] require modifying both the ontology and the reasoner).

5 Conclusions and Future Work

We introduce an ontology model capable of handling temporal information in ontologies. The proposed model extends the 4D fluent representation of [4] to handle both quantitative and qualitative temporal information. The representation mechanism incorporates reasoning rules for inferring certain temporal relations from existing ones and for checking temporal assertions for consistency. Extending the model to support spatial relations and addressing scalability issues using appropriate indexing mechanisms are directions for further research.

Extending TOQL [4] to handle the proposed 4D fluent representation is another contribution of this work. A desirable feature of TOQL is that it does not require that the user be familiar with the peculiarities of the underlying 4D fluent representation mechanism (which may be complicated leading to complicated query expressions in other query languages such as SPARQL [8]). Extending SPARQL, the current W3C standard to support 4D fluents and similar operators is an important issue for future research. t-SPARQL [18] is an example of work along these lines. Notice though that t-SPARQL suggest using named graphs as the underlying temporal representation (does not support 4D fluents) and therefore, does not maintain full OWL expressiveness and has no reasoning support.

References

1. Grenon, P., Smith, B.: SNAP and SPAN: Towards Dynamic Spatial Ontology. Spatial Cognition and Computation 4(1), 69–104 (2004)
2. Noy, N., Rector, B.: Defining N-ary Relations on the Semantic Web. W3C Working Group Note 12 (April 2006), http://www.w3.org/TR/swbp-n-aryRelations/
3. Aduna, B.V.: The SeRQL query language. User Guide for Sesame 2.1, ch. 9 (2002–2008), http://www.openrdf.org/doc/sesame2/2.1.2/users/ch09.html
4. Baratis, E., Petrakis, E.G.M., Batsakis, S., Maris, N., Papadakis, N.: TOQL: Temporal Ontology Querying Language. In: Mamoulis, N., Seidl, T., Pedersen, T.B., Torp, K., Assent, I. (eds.) SSTD 2009. LNCS, vol. 5644, pp. 338–354. Springer, Heidelberg (2009)
5. Hobbs, J.R., Fang, P.: Time Ontology in OWL. W3C Recommendation (September 2006), http://www.w3.org/TR/owl-time/
6. Klein, M., Fensel, D.: Ontology Versioning for the Semantic Web. In: International Semantic Web Working Symposium (SWWS 2001), California, USA, July–August 2001, pp. 75–92 (2001)
7. McGuinness, D.L., VanHarmelen, F.: OWL Web Ontology Language Overview. W3C Recommendation (February 2004), http://www.w3.org/TR/owl-features
8. Prud'hommeaux, E., Seaborne, A.: SPARQL Query Language for RDF. W3C Recommendation (January 2008), http://www.w3.org/TR/rdf-sparql-query
9. Welty, C., Fikes, R.: A Reusable Ontology for Fluents in OWL. Frontiers in Artificial Intelligence and Applications 150, 226–236 (2006)
10. Gregersen, H., Jensen, C.S.: Temporal Entity Relationship Models - A Survey. IEEE Transactions on Knowledge and Data Engineering 3, 464–497 (1999)
11. Artale, A., Franconi, E.: A Survey of Temporal Extensions of Description Logics. Annals of Mathematics and Artificial Intelligence 30(1-4) (2001)
12. Lutz, C., Wolter, F., Zakharyaschev, M.: Temporal Description Logics: A Survey. In: Proc. TIME 2008. IEEE Press, Los Alamitos (2008)
13. Gutierrez, C., Hurtado, C., Vaisman, A.: Introducing Time into RDF. IEEE Transactions on Knowledge and Data Engineering 19(2), 207–218 (2007)
14. Renz, J., Nebel, B.: Qualitative Spatial Reasoning using Constraint Calculi. In: Handbook of Spatial Logics, pp. 161–215. Springer, Netherlands (2007)
15. Sheth, A., Arpinar, I., Perry, M., Hakimpour, F.: Geospatial and Temporal Semantic Analytics. In: Karimi, H.A. (ed.) Handbook of Research in Geoinformatics, ch. XXI (2009)

16. Parsia, B., Sirin, E.: Pellet: An OWL DL reasoner. In: McIlraith, S.A., Plexousakis, D., van Harmelen, F. (eds.) ISWC 2004. LNCS, vol. 3298. Springer, Heidelberg (2004)
17. Allen, J.F.: Maintaining Knowledge About Temporal Intervals. Communications of the ACM 26, 832–843 (1983)
18. Tappolet, J., Bernstein, A.: Applied Temporal RDF: Efficient Temporal Querying of RDF Data with SPARQL. In: Aroyo, L., Traverso, P., Ciravegna, F., Cimiano, P., Heath, T., Hyvönen, E., Mizoguchi, R., Oren, E., Sabou, M., Simperl, E. (eds.) ESWC 2009. LNCS, vol. 5554, pp. 308–322. Springer, Heidelberg (2009)
19. Chen, H., Perich, F., Finin, T., Joshi, A.: SOUPA: Standard Ontology for Ubiquitous and Pervasive Applications. In: Int. Conference on Mobile and Ubiquitous Systems: Networking and Services, pp. 258–267 (2004)
20. Stocker, M., Sirin, E.: PelletSpatial: A Hybrid RCC-8 and RDF/OWL Reasoning and Query Engine. OWLED (2009)
21. Horrocks, I., Kutz, O., Sattler, U.: The Even More Irresistible SROIQ. In: Proc. KR 2006, Lake District, UK (2006)
22. Grau, B.C., Horrocks, I., Motik, B., Parsia, B., Patel-Schneider, P., Sattler, U.: OWL 2: The Next Step for OWL. In: Web Semantics: Science, Services and Agents on the World Wide Web, vol. 6, pp. 309–322 (2008)
23. Milea, V., Frasincar, F., Kaymak, U.: Knowledge Engineering in a Temporal Semantic Web Context. In: The Eighth International Conference on Web Engineering (ICWE 2008). IEEE Computer Society Press, Los Alamitos (2008)
24. Motik, B., Patel-Schneider, P.F., Horrocks, I.: OWL 2 Web Ontology Language: Structural Specification and Functional-Style Syntax. W3C Recommendation (2009), http://www.w3.org/TR/owl2-syntax/
25. Nebel, B., Burckert, H.J.: Reasoning about Temporal Relations: A Maximal Tractable Subclass of Allen's Interval Algebra. Journal of the ACM (JACM) 42(1), 43–66 (1995)
26. de Bruijn, J.: RIF RDF and OWL Compatibility. W3C Working Draft (July 2009), http://www.w3.org/TR/rif-rdf-owl/
27. Horrocks, I., Patel-Schneider, P.F., Boley, H., Tabet, S., Grosof, B., Dean, M.: SWRL: A Semantic Web Rule Language Combining OWL and RuleML. W3C Member submission (2004), http://www.w3.org/Submission/SWRL/
28. van Beek, P., Cohen, R.: Exact and approximate reasoning about temporal relations. Computational intelligence 6(3), 132–147 (1990)
29. Vilain, M., Kautz, H., van Beek, P.: Constraint propagation algorithms for temporal reasoning: a revised report. In: Weld, D.S., de Kleer, J. (eds.) Readings in qualitative reasoning about physical systems, pp. 373–381 (1989)
30. Krokhin, A., Jeavons, P., Jonsson, P.: Reasoning about temporal relations: The tractable subalgebras of Allen's interval algebra. Journal of the ACM 50(5), 591–640 (2003)
31. Motik, B., Sattler, U., Studer, R.: Query Answering for OWL-DL with rules. In: Web Semantics: Science, Services and Agents on the World Wide Web, Rules Systems, vol. 3(1), pp. 41–60 (July 2005)
32. Kolovski, V., Parsia, B., Sirin, E.: Extending the SHOIQ (D) tableaux with dl-safe rules: First results. In: Proceedings International Workshop on Description Logic (DL 2006) (2006)
33. Batsakis, S., Petrakis, E.: SOWL:Spatio-temporal Representation, Reasoning and Querying over the Semantic Web. In: 6th International Conference on Semantic Systems (I-SEMANTICS 2010), Graz, Austria, September 1-3 (2010)

Combining a Multi-Document Update Summarization System –CBSEAS– with a Genetic Algorithm

Aurélien Bossard and Christophe Rodrigues

Abstract. In this paper, we present a combination of a multi-document summarization system with a genetic algorithm. We first introduce a novel approach for automatic summarization. CBSEAS, the system which implements this approach, integrates a new method to detect redundancy at its very core in order to produce summaries with a good informational diversity. However, the evaluation of our system at TAC 2008 —Text Analysis Conference— revealed that system adaptation to a specific domain is fundamental to obtain summaries of an acceptable quality.

The second part of this paper is dedicated to a genetic algorithm which aims to adapt our system to specific domains. We present its evaluation by TAC 2009 on a newswire articles summarization task and show that this optimization is having a great influence on both human and automatic evaluations.

1 Introduction

As more information becomes available online, people confront a new problem: disorientation due to the abundance of information. Document retrieval and text summarization systems can be used to address this problem. While document retrieval engines can help a user to filter out documents, summarization systems can extract and present the essential content of these documents.

Recently, the DUC —Document Understanding Conference— now known as TAC —Text Analysis Conference[1]— evaluation campaigns have proposed to evaluate automatic summarization systems. These competitions have led to recent improvements in summarization and its evaluation.

Aurélien Bossard · Christophe Rodrigues
Laboratoire d'informatique de Paris Nord, CNRS UMR 7030
Université Paris 13, 93430 Villetaneuse, France
e-mail: firstname.lastname@lipn.univ-paris13.fr

[1] http://nist.tac.gov

I. Hatzilygeroudis and J. Prentzas (Eds.): Comb. of Intell. Methods and Appl., SIST 8, pp. 71–87.
springerlink.com

In this paper, we present our system, called CBSEAS —Clustering Based Sentence Extractor for Automatic Summarization— and its adaptation to the newswire article summarization task: the use of a genetic algorithm which aims at finding automatically the best suited parameter combination as input of the system.

We first give a quick overview of existing automatic summarization systems. In a second section, we describe our system. We then present our method for parameters optimization, based on a genetic algorithm. In a last section, we discuss the results obtained by our system: its performance on the summarization task, and the influence of the parameters values.

2 Automatic Extractive Summarization Overview

The extractive approaches to automatic summarization consist in selecting the most pertinent sentences or phrases and assemble them together to create a summary. This section gives an overview of this kind of approaches.

2.1 Feature-Based Approaches

Edmundson [7] defined textual clues which can be used to determine the importance of a sentence. In particular, he set a list of cue words, such as "hardly" or "impossible", using term frequency, sentence position (in a news article for example, the first sentences are the most important) and the number of words occuring in the title. These clues are still used by recent systems, like the one of Kupiec [12].

This kind of approaches does not take into account the overall content of the documents. That is why automatic summarization has evolved into sentence selection using the "centrality" feature: the sentence importance relatively to the overall documents content.

2.2 Centrality-Based Approaches

Other systems focus on term frequency. Luhn [15] led the way of frequency-based sentence extraction systems. He proposed to build a list of important terms. The importance of a term depends on wether or not its frequency belongs or not to a predefined range. The more a sentence presents words belonging to this list, the more important it is. Radev [19] took advantage of the advances in text statistics by integrating the tf.idf metric to Luhn's method. The list of important terms, that Radev calls "centroid", is composed of the n terms with the highest *tf.idf* –the *tf.idf* metric was introduced by Salton[20]. The sentences are ranked according to their similarity to the centroid. Radev also included a post-processing step to eliminate redundancy from the summary. He implemented this method in an online multi-document summarizer, MEAD[2] [18].

Radev further improved MEAD using another sentence selection method which he named "Graph-based centrality" [8]. It consists in computing similarity between

[2] http://www.newsinessence.com/clair/meaddemo/demo.cgi

sentences, and then selecting sentences which are considered as "central" in a graph where nodes are sentences and edges are similarities. The most central sentences are those which have been visited most after a random walk on the graph. This method is inspired by the concept of *prestige* in social network.

The clue-based, term frequency-based and "graph-based centrality" methods are efficient when selecting the sentences which reflect the global content of the documents to be summed up. Such a sentence is called "central". However, these methods are not designed to generate good summaries according to informational diversity. Now, informational diversity is almost as important as centrality when evaluating a summary. Indeed, a summary should contain all the important pieces of information which should not be repeated.

2.3 Dealing with Diversity

In multi-document summarization, the risk of extracting two sentences conveying the same information is greater than in a single-document summarization problematic. Moreover, identifying redundancy is a critical task, as information appearing several times in different documents can be qualified as important.

The previously presented systems are dealing with redundancy as a post-processing step. Goldberg [9], assuming that redundancy should be the key concept of multi-document summarization, offered a method to deal with redundancy at the same time as sentence selection. For that purpose, he used a "Markov absorbing chain random walk" on a graph representing the different sentences of the corpus to summarize.

MMR-MD, introduced by Carbonnel in [5], is a measure which needs a passage clustering: all passages considered as synonyms are grouped into the same clusters. MMR-MD takes into account the similarity to a query, coverage of a passage (clusters that it belongs to), content in the passage, similarity to passages already selected for the summary, belonging to a cluster or to a document that has already contributed a passage to the summary.

The problem of this measure lies in the clustering method: in the literature, clustering is generally fulfilled using a threshold. If a passage has a similarity to a cluster centroid higher than a threshold, then it is added to this cluster. This makes it a supervised clustering method.

Considering that diversity is the main issue in multi-document summarization, we want our method to first deal with diversity, grouping sentences in clusters according to the information they convey. The diversity management has to be unsupervised in order to be adapted to every type of documents. Our method will then apply local centrality-based selection methods to extract one sentence per cluster.

3 CBSEAS: A Clustering-Based Sentence Extractor for Automatic Summarization

We want to specifically manage the multi-document aspect by considering redundancy as the main issue of multi-document summarization. Indeed, we consider the documents to summarize as made up by groups of sentences carrying the same information. In each of these clusters, one sentence can be considered as central. Extracting this sentence, and not another one, in every cluster can lead to summaries in which the risk of redundancy is minimized. The summaries generated with this method may carry a good informational diversity. We here briefly present our system, which is further described in [2].

3.1 Pre-processing

All sentences go through a POS tagger, TreeTagger[3]. While studying news corpora, we identified several categories of news. Only a few of them present some particularities which make them worthwhile for an automatic summarization system. Details are available in [4]. Documents are classified using a keywords/structure clue based categorizer, into four categories:

- Classic news (1: presentation of the event, 2: the premisses, possibly 3: the consequences or projection in the future);
- Chronologies (list of related events ordered chronologically, *cf* Figure 1);
- Comparative news (the state of the article topic in different places or at different times, *cf* Figure 1);
- Enumerative news (an enumeration of facts, recommandations...).

The last three categories are very interesting for an automatic summarizer. In fact, they make up at most 5% of the total number of newswire articles in AQUAINT-2[4]. But, in the training corpus of the "Update Task", they contain 80% of the pertinent information. Moreover, they are written in a concise style, and can be easily inserted into a summary.

$$sim(s_1, s_2) = \frac{\dfrac{\sum_{mt} weight(mt) \times fsim(s_1, s_2)}{fsim(s_1, s_2) + gsim(s_1, s_2)}}{\sum_{mt} weight(mt)} \quad (1)$$

$$fsim(s_1, s_2) = \sum_{n_1 \in s_1} \sum_{n_2 \in s_2} tsim(n_1, n_2) \times \frac{tfidf(n_1) + tfidf(n_2)}{2} \quad (2)$$

$$gsim(s_1, s_2) = card\left((n_1 \in s_1, n_2 \in s_2) \mid tsim(n_1, n_2) < \delta\right) \quad (3)$$

where mt are the morphological types, s_1 and s_2 the sentences, tsim the similarity between two terms using WordNet and the JCn similarity measure [11] and δ a similarity threshold.

[3] http://www.ims.uni-stuttgart.de/projekte/corplex/TreeTagger/

[4] AQUAINT-2 is a corpus built by NIST and composed of 900.000 news articles from different sources (AFP, APW, NYT...)

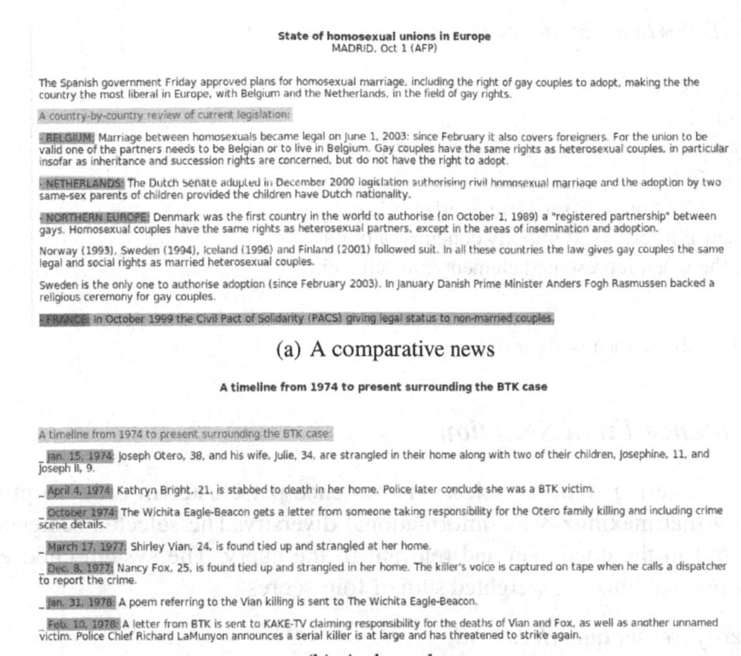
State of homosexual unions in Europe
MADRID, Oct 1 (AFP)

The Spanish government Friday approved plans for homosexual marriage, including the right of gay couples to adopt, making the the country the most liberal in Europe, with Belgium and the Netherlands, in the field of gay rights.

A country-by-country review of current legislation:

- BELGIUM: Marriage between homosexuals became legal on June 1, 2003: since February it also covers foreigners. For the union to be valid one of the partners needs to be Belgian or to live in Belgium. Gay couples have the same rights as heterosexual couples, in particular insofar as inheritance and succession rights are concerned, but do not have the right to adopt.

- NETHERLANDS: The Dutch Senate adopted in December 2000 legislation authorising civil homosexual marriage and the adoption by two same-sex parents of children provided the children have Dutch nationality.

- NORTHERN EUROPE: Denmark was the first country in the world to authorise (on October 1, 1989) a "registered partnership" between gays. Homosexual couples have the same rights as heterosexual partners, except in the areas of insemination and adoption.

Norway (1993), Sweden (1994), Iceland (1996) and Finland (2001) followed suit. In all these countries the law gives gay couples the same legal and social rights as married heterosexual couples.

Sweden is the only one to authorise adoption (since February 2003). In January Danish Prime Minister Anders Fogh Rasmussen backed a religious ceremony for gay couples.

- FRANCE: in October 1999 the Civil Pact of Solidarity (PACS) giving legal status to non-married couples.

(a) A comparative news

A timeline from 1974 to present surrounding the BTK case

A timeline from 1974 to present surrounding the BTK case:

_ Jan. 15, 1974: Joseph Otero, 38, and his wife, Julie, 34, are strangled in their home along with two of their children, Josephine, 11, and Joseph II, 9.

_ April 4, 1974: Kathryn Bright, 21, is stabbed to death in her home. Police later conclude she was a BTK victim.

_ October 1974: The Wichita Eagle-Beacon gets a letter from someone taking responsibility for the Otero family killing and including crime scene details.

_ March 17, 1977: Shirley Vian, 24, is found tied up and strangled at her home.

_ Dec. 8, 1977: Nancy Fox, 25, is found tied up and strangled in her home. The killer's voice is captured on tape when he calls a dispatcher to report the crime.

_ Jan. 31, 1978: A poem referring to the Vian killing is sent to The Wichita Eagle-Beacon.

_ Feb. 10, 1978: A letter from BTK is sent to KAKE-TV claiming responsibility for the deaths of Vian and Fox, as well as another unnamed victim. Police Chief Richard LaMunyon announces a serial killer is at large and has threatened to strike again.

(b) A chronology

Fig. 1 News examples

3.2 Sentence Pre-selection

First, our system ranks all the sentences according to their similarity to the documents centroid, composed of the m terms with the highest *tf.idf*. In the case a user query is provided, the sentences are ranked according to their relevance to the query. We then select the best ranked sentences, using an empiric threshold. This method has been changed with the integration of the genetic algorithm, as shown in Sec. 4.

3.3 Sentence Clustering

Similarity between sentences is computed using a variant of the "Jaccard" measure, shown in Equations 1, 2 and 3. Other similarity measures exist, such as cosine similarity, but this measure allows us to take into account the similarity between two different terms in the sentence similarity computation. This point is important as linguistic variation could otherwise not be managed.

Once the similarities are computed, we cluster the sentences using fast global k-means (description of the algorithm is in Figure 2) using the similarity matrix.

for all $e_j inE$ *%%Initialize the first cluster with all the elements*
$C_1 \leftarrow e_j$
for i from 1 to k do
for j from 1 to i
center$(C_j) \leftarrow argmax_{e_m} \sum_{e_n \in C_j} sim(e_m, e_n)$
for all e_j in E
$e_j \rightarrow C_l | C_l maximizessim(center(C_l, e_j)$
add a new cluster: C_i. It initially contains only its
center, the worst represented element in its cluster.
done

Fig. 2 Fast global k-means algorithm

3.4 Sentence Final Selection

After this clustering step, we select one sentence per cluster in order to produce a summary that maximizes the informational diversity. The selected sentence has to be central in the document and relevant to the query. The system chooses the sentence that maximizes a weighted sum of four scores :

- Similarity to user query/*centroid*;
- Similarity to cluster center;
- Important sentence score (implemented after TAC 2008 campaign);
- Difference in length between the scored sentence and the desired sentence length.

The "Important sentence score" is the inverse of the sentence position in the document if the sentence is part of a "classic news", or 1 if the sentence is part of the body of a news classified as a chronology, an enumerative news or a comparative news.

3.5 Managing Update for TAC "Update Task"

Sometimes, a user wants to know what is new about a topic since the last time he has read news about it. That is why the TAC 2008 and TAC 2009 "Update Task" consisted in summarizing a first document set, then summarizing what is new in a second document set.

CBSEAS –Clustering-Based Sentence Extractor for Automatic Summarization– clusters semantically close sentences. In others terms, it creates different clusters for semantically distant sentences. Our clustering method can also be used to differenciate sentences carrying new pieces of information from sentences carrying already known pieces of information, and so for managing update. In fact, sentences carrying old pieces of information are semantically close from the sentences that a user has already read.

CBSEAS has proven to be efficient at grouping together semantically close sentences and differentiate semantically far ones. In fact, the results obtained by CBSEAS on TAC 2008 Opinion Task are good, as CBSEAS appears at the third place for avoiding redundancy in the summaries [3]. This is another reason for using our clustering method to differentiate update sentences from non-update ones.

Before trying to identify update sentences, we need to modelize the pieces of information that the user requesting the update summary has already read. We can then confront the new documents to this model in order to determine if sentences from these documents carry new pieces of information. So the first step of our algorithm is to cluster the sentences from the documents the user has already read –which we call D_I– into k_I groups, as in Sec. 3.3 for the generation of a standard summary.

The model thus computed –M_I– is then used for the second step of our algorithm, which consists in determining if a sentence from the new documents –D_U– is to be grouped with the sentences from D_I, or to create a new cluster which will only contain update sentences. *Fast global k-means* algorithm, slightly modified, can be used to confront elements to a previously established model in order to determine if these elements can be an integral part of the model. We here describe the second clustering part of our update algorithm.

First, our algorithm selects the sentences from D_U same as for D_I (*cf* Sec. 3.2). Then, it computes the similarities between sentences from D_U with the cluster centers of M_I and between all the sentences from D_U. Then it adds the new sentences to M_I, and iterates *fast global k-means* from the k_I iteration with the following constraints:

- The sentences from D_I can not be moved to another cluster; this is done to preserve the M_I model which encodes the old pieces of information. It also avoids to disturb the semantic range of the new clusters that bear novelty.
- The cluster centers from M_I can not be recomputed; as the semantic range of a cluster depends directly on its center, this prevents the semantic range of M_I clusters from being changed by the integration of new elements from D_U.

In order to favor sentences from the second set of document being part of the update clusters, a negative weight can be assigned to the similarities between sentences belonging to the first document set and sentences belonging to the second.

Once the update clusters have been populated, the update summary is generated by extracting one sentence per update cluster, as in Sec. 3.4.

4 Optimizing CBSEAS Parameters

News article summarization differs from scientific article summarization or technical report summarization. When aiming at finding similar sentences in order to detect central sentences in a technical report, a system should not focus on the same markers as for blogs or novel summarization. Dealing with scientific articles, centrality could not be the best indicator of sentence importance. Teufel has shown in [21] that examining the rhetorical status of a sentence —its position in the document structure, if it contains cue phrases...— is a good way to figure out if it should appear in the final summary.

Our participation to both the "Update Task" (*cf* Sec. 3.5) and the "Opinion Task" —Summarizing opinions found in blogs— of TAC 2008 showed us that our system can be competitive; it ranked second on the "Opinion Task", but its poor behavior on the "Update Task" showed that adaptation Splays a crucial role in performing better on this task. For this purpose, we have first implemented a score that takes into account specific news structure traits (*cf* Sec. 3.4), and have chosen to use a learning technique that automatically adapts CBSEAS' weights according to a scoring method.

TAC 2008 campaign provided us a corpus, manual reference summaries, and an automatic evaluation framework: ROUGE[5]. ROUGE is a package of automatic evaluation measures using unigram co-occurrences between summary pairs [13]. When computing ROUGE scores between an automatic summary and one or more manual summary, we can efficiently evaluate the information content of the automatic summary. Also, our system takes fourteen parameters as input:

1. number of sentences desired as output;
2. average desired sentence length ;
3. weights of proper names, (4.) nouns, (5.) adjectives, (6.) adverbs, (7.) verbs and (8.) numbers in the similarity function (*cf* Sec. 3.3);
9. number of pre-selected sentences from the first and the (10.) second document sets ;
11. weight of similarity to cluster center, (12.) important sentence score, (13.) and length difference in the final sentence selection scoring (*cf* Sec. 3.4);
14. reduction of similarities between first document set and second document set sentences (*cf* Sec. 3.4).

We have all it takes for an environment interactive learning method.

4.1 Overview of Parameters Optimization for Automatic Summarization

In the field of trainable summarizers, systems combine basic features and try to find the best weight combination using an algorithm that adapts weights to maximize a fitness score. Kupiec [12] and Aone [1] used similar features to Edmundson [7] and optimized the weight of every feature using a trainable feature combiner using Bayesian network. MCBA [23] added two scores: a centrality score —intersection of sentence keywords and the other sentences keywords on the union of sentence keywords and the other sentences keywords)— and the similarity to title. The best weight combination is approximated using a genetic algorithm. Osborne used a gradient search method to optimize the feature weights[17].

In a more statistical-oriented approach, the PYTHY system [22] used standard features and different frequency-based features. The search for the best weight combination was based on a dynamic programming solution for the knapsack problem described in [16].

---
[5] http://berouge.com

4.2 *What Type of Algorithm?*

In our case, we cannot prove the regularity and continuity of a function from the hypothesis space to the summary score. Indeed, the parameters we use are not only weights for linear features combination. Now, function continuity is a pre-required for gradient search methods to work correctly. Moreover, as some parameters operate at different steps of our algorithm and on different aspects of sentence selection, building up a probabilistic model of hypothesis space that takes into account parameters dependencies is too complicated. The number of parameters (14) emphasizes the hugeness of the search space. Consequently, a genetic algorithm seems an appropriate method to learn the best parameters combination.

Genetic algorithms have been introduced by John Holland [10]. Holland aims at using species natural adaptation metaphor in order to automatically realize an optimal adaptation to an environment. The main idea is to generate individuals, and by means of mutation and crossing over selected individuals, to father a new generation of individual that will be more adapted to its environment than the previous one.

4.3 *ROUGE-SU4 Metric Liability*

We are using ROUGE-SU4 metric to automatically evaluate the quality of the summaries. We won't describe this metric, but one can find details about it in [13]. The liability of this metric is crucial for the genetic algorithm. During TAC 2008 campaign, three evaluations have been conducted:

- an entirely manual evaluation: assessors had to fill a grid with scores such as non-redundancy, readability, overall responsiveness[6], grammaticality, readability;
- pyramid evaluation [14], which consists in manually comparing the information available in the automatic summaries with the information available in the reference summaries;
- ROUGE evaluation.

Amongst the ten best ranked systems in responsiveness score, only four appeared in the top ten of ROUGE-SU4 scores. However, five out of the six other systems from this top ten ranked between the average and the poorest system in readability. This means that readability has a great influence on a human assessor judging the responsiveness. We noticed that systems ranked low in readability were using rewriting rules or sentence compression methods that make summaries less readable. Here is an extract of a summary created by one of these systems: *"The A380 will take over from the Boeing 747 (....?). The Airbus official said he had not seen any sign (of what?). Airbus says the A380 will produce half (as what?) as the 747. Most airports originally thought to accommodate (....?) the A380. The A380 is designed to carry 555 passengers. The plane's engineers will begin to find out (what?)."*.

[6] Overall responsiveness is the answer to the question : "How much would you pay for this summary?"

One can see that this summary, although it obtained good ROUGE scores, is not understandable. The summarization system has removed phrases that are essential for sentences comprehension.

ROUGE-SU4 is a good metric to evaluate different summaries created by extraction systems that do not modify extracted sentences when summarizing documents such as newswire articles, where sentences are all syntactically correct. So this metric is adapted to our optimization problem.

4.4 Our Genetic Algorithm

4.4.1 The Individuals

Each individual is composed of 14 parameters, which are described in Section 4. We empirically set their variation space. The Table 1 shows the space in which they fluctuate.

4.4.2 Individuals Selection Method

The evaluation of one individual is for us a time costly operation. That is the reason why we have chosen a tournament selection method, which has the advantage to be easily parallelized. For each generation of γ individuals, μ tournaments between λ individuals are organized. The winner of each tournament is selected to be part of the next generation parents. Another advantage of this method lies in the fact that it preserves diversity because the selected individuals are not forced to be the best ones. This prevents the algorithm from getting stuck in a local minimum.

$$\delta_i = \begin{cases} \lceil log(val_i - min_i) \times rand(0,1) \rceil, & val_i \neq min_i, rand_i(0,1) < lower_i \quad (4) \\ 1, & val_i = min_i, rand_i(0,1) < lower_i \quad (5) \\ \lceil log(val_i - max_i) \times rand(0,1) \rceil, & val_i \neq max_i, rand_i(0,1) > lower_i \quad (6) \\ 1, & val_i = max_i, rand_i(0,1) > lower_i . \quad (7) \end{cases}$$

where val_i is the value of parameter i,
and

$$lower_i = \frac{val_i - min_i}{max_i - min_i}, \tag{8}$$

with i from 1 to 14.

Table 1 Parameters' variation space

parameter	min	max	step
num. of sentences	1	20	1
av. length	1	20	1
num. of pre-selected sent.	1	200	1
num. of pre-selected sent. update	1	200	1
nouns weight	1	300	1
proper names weight	1	300	1
verbs weight	1	300	1
adjectives weight	1	300	1
adverbs weight	1	300	1
numbers weight	1	300	1
cluster center sim weight	1	300	1
important sent. score weight	1	300	1
length difference score weight	1	300	1
update sim reduction	0	1	0.01

4.4.3 Mutation Operator

As we do not know what parameters are dependent one to another, we want to change several parameters at the same time. In order to avoid a too heavy variation due to the simultaneous mutation of several parameters, we have chosen to limit the variation quantity (δ_i) of a parameter, weakening the probability to obtain a strong variation. We do that by using a logarithmic variation described in Equations 4 and 8.

4.4.4 Creating a New Generation

Each generation is composed of 100 individuals. The algorithm organizes twenty tournaments with fifteen randomly selected representatives. This seems to be a good compromise between quick evolution and diversity preservation. Each new generation is composed of the twenty winners, forty individuals created by mutating the winners, and the last forty created by randomly crossing the winners.

4.5 Training and Evaluation Data

TAC 2008 and 2009 "Update Task" consisted in creating two abstracts for forty-eight pairs of document sets. As computing a summary is time expensive, we decided to limit the training data to nine pairs of document sets. The evaluation data is composed of the forty other pairs of document sets.

5 Evaluation

TAC 2008 campaign has shown that automatic evaluation was still not as trustable as manual evaluation when dealing with summaries [6]. Although automatic evaluation proves to be useful to quickly judge the quality of a summary or to act as a fitness score for a learning algorithm, we cannot entirely rely on automatic evaluation. Our goal is to figure out at what point the optimization of the parameters really improves the quality of the automatic created summaries. We propose here two ways to do this: using ROUGE scores to see if the optimized parameters have led to an enhancement on the evaluation data, and letting an assessor judge if there is a visible improvement of the summaries quality.

We selected the best manually evaluated summarizer from TAC 2008, and our summarizer CBSEAS before and after the optimization. We selected fifteen pairs of document sets, and submitted the results of both of the three systems to an assessor, giving the automatically created summaries random ids, in order to avoid the assessor being able to identify the origin of summaries.

We then asked two questions to the assessor:

- Which one of the three summaries reflects best the documents content? (this summary gets the score 6)
- Compared to the best summary, give a score between 1 and five to the two other ones:
  - 5: the summary is almost as informative as the best one;
  - 4: the summary is a bit less informative than the best one;
  - 3: the summary is less informative than the best one;
  - 2: the summary is really less informative than the best one;
  - 1: no comparison is possible, the best summary overtakes this one.

We participated to TAC 2009 in order to validate that our system is performing better and to evaluate its competitiveness.

6 Results and Discussion

The Table 2 shows the combination of features selected by the genetic algorithm after 80 generations. It points out that setting a low weight of the proper names weight has a positive influence on the summary ROUGE scores. Also, the more important types seem to be the common names, adjectives and verbs. Adverbs are having a lesser influence on the summary quality.

The weight of proper names is so small because most of the selected sentences contain the same proper names, due to the fact that pre-selected sentences are close to the user query. This query is indeed most of the time oriented by named entities. So, having proper names playing an important role in sentence similarity computation brings noise to the similarity measure and affects negatively the clustering algorithm. In a more general way, this validates the observation of Aone et al. [1]: decreasing the impact of proper names in the sentence selection method for automatic news summarization increases the quality of the summaries.

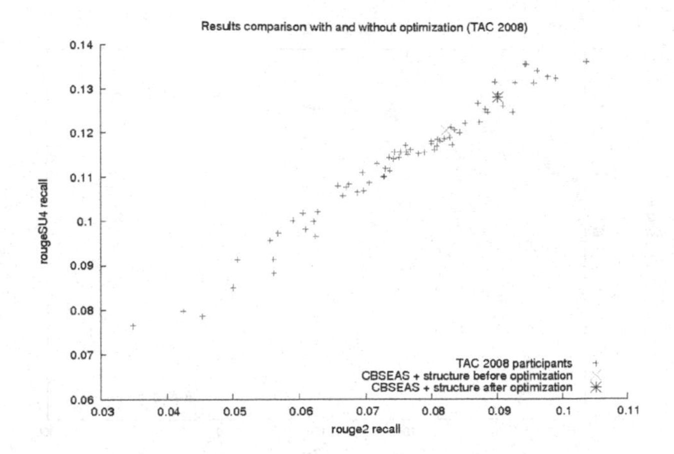

Fig. 3 ROUGE scores comparison of CBSEAS with TAC 2008 other participants

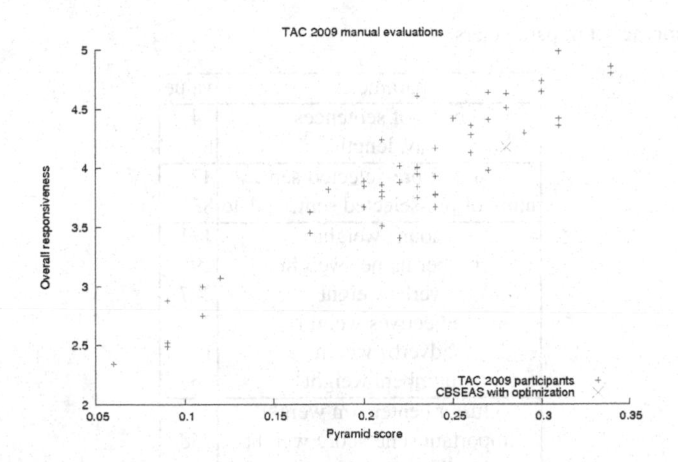

Fig. 4 ROUGE scores comparison of CBSEAS with TAC 2009 other participants

Setting the variable "update sim reduction" in a way that strenghtens the similarities between sentences from the first and the second set of documents leads to the generation of higher scored summaries. This means that decreasing the probability that a sentence from the second document set will appear in an update cluster improves the quality of the update management.

It is interesting to note that the feature "similarity to cluster center" gets the lowest weight in the last step of our algorithm. As recent works have proven the pertinence of graph-based methods for automatic summarization, this tends to prove that our similarity score is not adapted to such a feature. Other similarity measures should be reassessed in order to increase the impact of this feature.

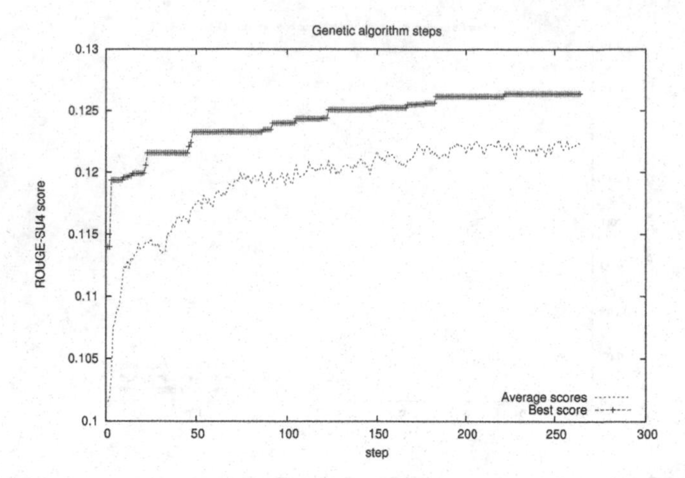

Fig. 5 Average of individual scores, and best individual for each generation

Table 2 Winning set of parameters

parameter	value
num. of sentences	14
av. length	8
num. of pre-selected sent.	47
num. of pre-selected sent. update	83
nouns weight	171
proper names weight	29
verbs weight	207
adjectives weight	270
adverbs weight	12
numbers weight	66
cluster center sim weight	7
important sent. score weight	258
length difference score weight	72
update sim reduction	0.87

We observe that manual evaluation presented in Table 3 and automatic evaluation agree: optimizing our parameters for this task has led to an important improvement of the summaries quality, but CBSEAS still does not overtake the best automatic systems of TAC 2008. This has been confirmed by our participation to TAC 2009 and the manual results of this conference, as shown by Fig. 4 (Pyramid and overall responsiveness evaluations). However, the system ranks among the best quarter of all participating systems.

Table 3 Manual evaluation

	Best TAC system	CBSEAS w/o optimization	Optimized CBSEAS
Standard summaries			
Number of times winning	9	2	4
non winning summaries average score	4.7	3.9	4.3
Update summaries			
Number of times winning	8	2	5
non winning summaries average score	5	3.7	4.5
Overall scores			
Number of times winning	17	4	9
non winning summaries average score	4.8	3.8	4.4

7 Conclusion

In this article, we presented our approach to generic multi-document summarization and update management, and the integration of news articles structure to our system, CBSEAS. We also presented a way to optimize the system we have developed via a genetic algorithm. The results obtained by both manual and automatic evaluations have shown us that the quality of our summaries has greatly improved. The impact of domain characteristics are important when automatically summarizing documents. The use of a genetic algorithm to optimize the features treatment in our systems has revealed some counter-intuitive observations. Although a human judgment is necessary, we cannot exclude automatic ways to find the best parameters combination for a given task. The results of TAC 2009 also show that our system still needs some improvements to rank among the very best systems. More linguistic methods, such as sentence compression or sentence reranking should be investigated to improve the overall quality of the summaries generated by CBSEAS.

Acknowledgement

Special thanks to Thibault Mondary and the GipiLab for having accepted to spend some time evaluating manually our work.

References

1. Aone, C., Okurowski, M.E., Gorlinsky, J.: Trainable, scalable summarization using robust nlp and machine learning. In: Proceedings of the 17th international conference on Computational linguistics, pp. 62–66. Association for Computational Linguistics, Morristown (1998), doi: http://dx.doi.org/10.3115/980845.980856
2. Bossard, A.: CBSEAS, a new approach to automatic summarization. In: Proceedings of the SIGIR 2009 Conference - Doctoral Consortium, Boston, USA (2009)
3. Bossard, A., Généreux, M., Poibeau, T.: Description of the LIPN System at TAC 2008: Summarizing Information and Opinions. In: Proceedings of the 2008 Text Analysis Conference, TAC 2008 Gaithersburg, United States, pp. 282–291 (2008), http://hal.archives-ouvertes.fr/hal-00397010/en/
4. Bossard, A., Poibeau, T.: Integrating document structure to an automatic summarizer. In: RANLP 2009, Borovets, Bulgaria (2009)
5. Carbonell, J., Goldstein, J.: The use of mmr, diversity-based reranking for reordering documents and producing summaries. In: SIGIR 1998: Proceedings of the 21st annual international ACM SIGIR conference, pp. 335–336. ACM, New York (1998)
6. Dang, H.T., Owczarzak, K.: Overview of the TAC 2008 update summarization task (DRAFT). In: Notebook papers and results of TAC 2008, Gaithersburg, Maryland, USA, pp. 10–23 (2008)
7. Edmundson, H.P., Wyllys, R.E.: Automatic abstracting and indexing—survey and recommendations. Commun. ACM 4(5), 226–234 (1961)
8. Erkan, G., Radev, D.R.: Lexrank: Graph-based centrality as salience in text summarization. Journal of Artificial Intelligence Research, JAIR (2004)
9. Goldberg, A.: Cs838-1 advanced nlp: Automatic summarization (2007), http://www.avglab.com/andrew/
10. Holland, J.H.: Adaptation in natural and artificial systems: An introductory analysis with applications to biology, control, and artificial intelligence. University of Michigan Press, Ann Arbor (1975)
11. Jiang, J.J., Conrath, D.W.: Semantic similarity based on corpus statistics and lexical taxonomy. In: International Conference Research on Computational Linguistics (ROCLING X), September 1997, p. 9008 (1997), http://adsabs.harvard.edu/cgi-bin/nph-bib_query?bibcode=1997cmp.lg....9008J
12. Kupiec, J., Pedersen, J., Chen, F.: A trainable document summarizer. In: SIGIR 1995: Proceedings of the 18th annual international ACM SIGIR conference on Research and development in information retrieval, pp. 68–73. ACM, New York (1995), doi: http://doi.acm.org/10.1145/215206.215333
13. Lin, C.Y.: Rouge: a package for automatic evaluation of summaries. In: Proceedings of the Workshop on Text Summarization Branches Out (WAS 2004), Barcelona, Spain (2004)
14. Lin, C.Y., Cao, G., Gao, J., Nie, J.Y.: An information-theoretic approach to automatic evaluation of summaries. In: Proceedings of the main conference on HLTC NACACL, pp. 463–470. Association for Computational Linguistics, Morristown (2006)
15. Luhn, H.P.: The automatic creation of literature abstracts. IBM Journal 2(2), 159–165 (1958)
16. McDonald, R.: A study of global inference algorithms in multi-document summarization. In: Amati, G., Carpineto, C., Romano, G. (eds.) ECiR 2007. LNCS, vol. 4425, pp. 557–564. Springer, Heidelberg (2007)

17. Osborne, M.: Using maximum entropy for sentence extraction. In: Proceedings of the ACL 2002 Workshop on Automatic Summarization, pp. 1–8. Association for Computational Linguistics, Morristown (2002), doi: http://dx.doi.org/10.3115/1118162.1118163
18. Radev, D., Allison, T., Blair-Goldensohn, S., Blitzer, J., Çelebi, A., Dimitrov, S., Drabek, E., Hakim, A., Lam, W., Liu, D., Otterbacher, J., Qi, H., Saggion, H., Teufel, S., Topper, M., Winkel, A., Zhu, Z.: MEAD - a platform for multidocument multilingual text summarization. In: Proceedings of LREC 2004, Lisbon, Portugal (2004)
19. Radev, D., Winkel, A.: Multi document centroid-based text summarization. In: ACL 2002 (2002)
20. Salton, G., McGill, M.J.: Introduction to modern information retrieval (1983)
21. Teufel, S., Moens, M.: Summarizing scientific articles - experiments with relevance and rhetorical status. Computational Linguistics 28 (2002)
22. Toutanova, K., Brockett, C., Gamon, M., Jagarlamudi, J., Hisami, S., Vanderwende, L.: The pythy summarization system: Microsoft research at DUC 2007. In: Proceedings of the HLT-NAACL Workshop on the Document Understanding Conference (DUC-2007), Rochester, USA (2007)
23. Yeh, J.Y., Ke, H.R., Yang, W.P.: Chinese text summarization using a trainable summarizer and latent semantic analysis. In: Lim, E.-p., Foo, S.S.-B., Khoo, C., Chen, H., Fox, E., Urs, S.R., Costantino, T. (eds.) ICADL 2002. LNCS, vol. 2555, pp. 76–87. Springer, Heidelberg (2002)

Extraction of Essential Events with Application to Damage Evaluation on Fuel Cells

Teppei Kitagawa, Ken-ichi Fukui, Kazuhisa Sato,
Junichiro Mizusaki, and Masayuki Numao

Abstract. Although sudden changes of the event phase in complex system may indicate underlying essential forces, such events are not frequent. In the present paper, we propose an essential event extractor (E^3) scheme to extract relatively rare but co-occurring event sequences in event phase transitions. In E^3, the self-organizing map (SOM) is used as vector quantization (VQ) to encode non-symbolic events and KeyGraph as a co-occurrence graph. Afterwards, event transitions on the co-occurrence graph can be obtained by referring to an occurrence density estimation on the topology map of VQ. We demonstrate the E^3 using an acoustic emission (AE) event sequence observed during a damage test of fuel cells and obtain reasonable and essential co-occurring damage sequences that exhibit mechanical effects.

1 Introduction

Most of the researches on mining from sequential or temporal data focuses on major trends or frequent patterns starting with Apriori[2, 10]. However, rare events play an important role for discovery of hidden forces under complex system, such as financial crash, rupture in a composite material, and earthquake. Sornette [26] stated, "Most complex systems in the natural and social sciences do exhibit rare and sudden transitions that occur over time intervals that are short compared with the characteristic time scales of their posterior evolution. Such extreme events express more than anything else the underlying forces usually hidden by an almost

Teppei Kitagawa · Ken-ichi Fukui · Msayuki Numao
The Institute of Scientific and Industrial Research, Osaka University,
8-1 Mihogaoka, Ibaraki, Osaka, 567-0047 Japan
e-mail: fukui@ai.sanken.osaka-u.ac.jp

Kazuhisa Sato · Junichiro Mizusaki
Institute of Multidisciplinary Research for Advanced Materials, Tohoku University,
2-1-1 Katahira, Aoba-ku, Sendai, 980-8577 Japan

I. Hatzilygeroudis and J. Prentzas (Eds.): Comb. of Intell. Methods and Appl., SIST 8, pp. 89–108.
springerlink.com

perfect balance and thus provide the potential for a better scientific understanding of complex systems." This work deals with an event sequence observed from such complex system, and we define *essential events* as follows:

Definition 1 (essential events). *Essential events are the events that exhibit the potential forces and appear in the phase transitions, which are short periods compared to their posterior evolution, by releasing the accumulated potential forces.*

Although there exist several researches on discovery of important rare events[29, 18, 28, 22, 1], these works deal with symbolic sequential data, which means each event is described by category. However, a lot of non-categorical data exists in the real world, such as financial data, sensor data. It is an important task to discover essential rare events from an event sequence where an event is described by a set of features or defined by (dis)similarity to other events.

We propose in this paper the essential event extractor (E^3) for a non-symbolic event sequence, using vector quantization (VQ) as an encoder. Then a co-occurrence analysis was applied to extract rare events which co-occur with fundamental high frequent events. The nature of VQ is to capture the entire data distribution by the small number of vectors, but not to divide the data distribution into meaningful clusters. This property is important for an unknown domain because most clustering algorithms have limitations, such as pre-setting of the number of clusters, a threshold to merge clusters, or a chaining effect[30].

We combined the self-organizing map (SOM)[12] as a VQ and KeyGraph[18] as a co-occurrence graph. The SOM and KeyGraph have very good compatibility because they are both forms of exploratory data analysis (EDA)[27], which support the user in investigating the data. As for a related work, Ohsawa applied KeyGraph to an earthquake sequence in order to discover risky active faults[19, 17]. However, each earthquake event is assigned to the nearest pre-defined active fault. The present paper contributes to the relaxation of this requirement, but requires (dis)similarity between events.

Based on an experiment using an acoustic emission (AE) event sequence obtained through an solid oxide fuel cells (SOFC) damage test, we demonstrate that the proposed E^3 can extract essential AE events. These AE events exhibit potential mechanical effects between the component materials of the fuel cells.

2 Essential Event Extractor (E^3)

2.1 Overview

The present work deals with a non-symbolic event sequence. Non-symbolic event in this work is defined as:

Definition 2 (non-symbolic event). *An event X_i is characterized by (dis)similarity to other events, i.e., $\forall j\ d(X_i,X_j)$. For example, $d(X_i,X_j)$ is defined by Euclidean distance of feature vectors between X_i and X_j.*

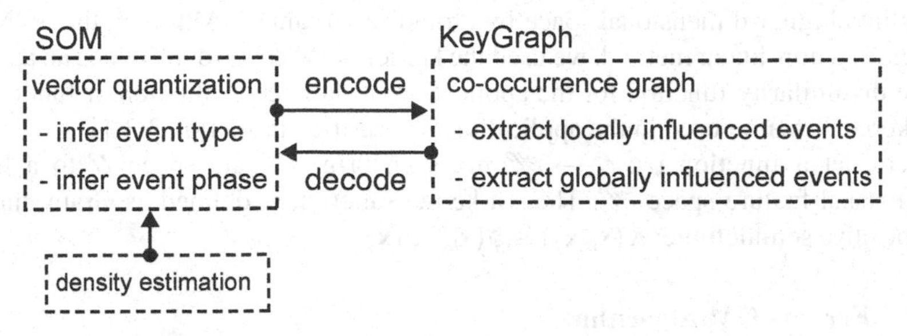

Fig. 1 Overview of essential event extractor (E^3)

Then, an event sequence is defined as:

Definition 3 (event sequence). *An event sequence is a set of ordered events denoted by* $\mathbf{D} = X_1, \ldots, X_T$*, where* X_t *refers to the* t^{th} *event.*

The overview of the proposed E^3 is illustrated in Fig. 1. In E^3, the SOM provides VQ as well as a low-dimensional representation of the data distribution, which allows the user to investigate individual events and to understand intuitively the entire picture of all events. In addition, the SOM encodes the entire data distribution by prototype vectors and provides codes to KeyGraph. KeyGraph then generates a graph that is based on the co-occurrence frequency of the prototypes within a certain period. KeyGraph extracts two types of events, the one is locally influenced (fundamental) events that are high occurrence frequency and the other is globally influenced events that are rare but co-occur with fundamental events. These globally influenced events are candidates of essential events.

Meanwhile, by estimating the occurrence probability density distribution of the prototypes within the topology map obtained by the SOM, the user can infer the event type and event phase based on the change in the estimated density distribution and the best matched events. Afterwards, referring to the transition direction by the estimated density change of the prototypes, the co-occurrence graph can be decomposed into small sequences that are highly correlated with phase transitions. Consequently, co-occurring event sequences in phase transitions, that is essential events, can be obtained within the low-dimensional map that exhibit the potential forces.

2.2 *Kernel SOM*

2.2.1 Overview

The SOM[12] is an unsupervised, competitive neural network learning model that has been applied in various domains, such as clustering and visualization of contents, control or monitoring of an industrial instrument, medical check, and so on[20]. The kernel SOM[4] was used in the present study, where the kernel trick is a method to extend a linear method to non-linear using a kernel function that maps into higher-dimensional space by an indirect manner. Although the SOM is originally a non-linear method, we used the kernel SOM so as to introduce an appropriate dissimilarity function for the application of damage evaluation of fuel cells. The kernel function used in the application is described in section 3.3.1

Here, let a function $\phi : \mathscr{O} \rightarrow \mathscr{H}$ maps an original data space $\mathscr{O}$ to a high dimensional feature space $\mathscr{H}$. Then, a kernel function is defined as Gram matrix of a positive semidefinite: $K(\mathbf{x}_i, \mathbf{x}_j) \equiv \phi(\mathbf{x}_i) \cdot \phi(\mathbf{x}_j)$.

2.2.2 Kernel SOM Algorithm

Suppose N input data $\{\mathbf{x}_1, \cdots, \mathbf{x}_N\}$ are given, where $\mathbf{x}_i = (x_{i,1}, \cdots, x_{i,v})$ is a v-dimensional data. Let M neurons of the prototype (reference) vectors be $\{\mathbf{m}_1, \cdots, \mathbf{m}_M\}$, where $\mathbf{m}_j = (m_{j,1}, \cdots, m_{j,v})$. In addition, let the position of M neurons in the topological layer be $\mathbf{r}_j = (x_j, y_j) : j = 1, \cdots, M$. The number of neurons and the layout of the topological layer must be pre-defined, and a regular or hexagonal grid is normally used. The following shows the learning algorithm that uses a batch process and decreasing strategy of the learning parameter.

S1 (Initialization). Initialize the prototype vectors $\{\mathbf{m}_1(t), \cdots, \mathbf{m}_M(t)\}$ randomly, also set an iteration counter as $t = 1$. In the kernel SOM, since $\mathbf{m}_i(t)$ cannot be calculated in $\mathscr{H}$, the dissimilarity between a prototype $\mathbf{m}_i(t)$ and an input $\mathbf{x}_n$ denoted by $\{d_{i,n}(t) : n = 1, \cdots, N\}$ is used instead of $\mathbf{m}_i(t)$.

S2 (Searching BMU). Search the best matching units (BMUs), in other words the winner neurons $\{c(\mathbf{x}_1), \cdots, c(\mathbf{x}_N)\}$ for all inputs by the nearest neuron:

$$c(n) = \arg \min_{i=1,\cdots,M} d_{i,n}(t). \tag{1}$$

S3 (Termination condition). Exit if the winner neurons $\{c(\mathbf{x}_1), \cdots, c(\mathbf{x}_N)\}$ were not changed or the iteration reached $t = t_{max}$.

S4 (Updating prototypes). Update the prototype vectors $\{\mathbf{m}_1(t), \cdots, \mathbf{m}_M(t)\}$ by the following equation:

$$\begin{aligned} d_{i,n}(t+1) &\equiv ||\phi(\mathbf{x}_n) - \mathbf{m}_i(t+1)||^2 \\ &= K(\mathbf{x}_n, \mathbf{x}_n) - 2\gamma \sum_j h_{c(j),i} K(\mathbf{x}_n, \mathbf{x}_j) \\ &\quad + \gamma^2 \sum_k \sum_l h_{c(k),i} h_{c(l),i} K(\mathbf{x}_k, \mathbf{x}_l), \end{aligned} \tag{2}$$

where $||\cdot||$ denotes L2-norm, and $\gamma = 1/\sum_n h_{c(n),i}$ is a normalization factor. In the kernel SOM, also $\phi(\mathbf{x}_n)$ cannot be calculated, the prototype vectors are updated in an indirect manner using the kernel function. In addition, $h_{i,j}$ is a neighborhood function that defines the effect of the neighborhood of the winner, and a Gaussian function is typically used as a neighborhood function:

$$h_{i,j} = \exp\Big(-\frac{\|\mathbf{r}_i - \mathbf{r}_j\|^2}{2\sigma^2}\Big). \tag{3}$$

S5 (Iterative processing). Decrease the neighborhood radius σ, also increase the iteration counter $t \leftarrow t+1$. Then, return to step S2.

2.3 *Density Estimation*

Density estimation is used to estimate the occurrence density of events on the topology map obtained by the SOM. Since a simple histogram has a problem with appropriate setting of the intervals to count the data, density estimation is an alternative to the histogram that estimates the generative density distribution at any point without setting of the intervals. Since no background knowledge is available with the data distribution of damage events of fuel cells, non-parametric density estimation is suitable. Therefore, we used kernel density estimation (KDE)[25].

The probability density by KDE at point $\mathbf{x} \in \mathscr{R}^v$ is given by:

$$P_{\mathrm{KDE}}(\mathbf{x}) = \frac{1}{Nb^v}\sum_{i=1}^{N} K\Big(\frac{\mathbf{x}-\mathbf{x}_i}{b}\Big), \tag{4}$$

where b is a band width, N is the number of data points, and v is the number of dimension. The larger b becomes, the smoother the distribution that can be obtained. In addition, $K(\mathbf{x})$ is a kernel function[1] at $\mathbf{x}$. The Gaussian kernel was used in the present study:

$$K(\mathbf{x}) = \frac{1}{2\pi^{-v/2}} \exp\Big(-\frac{\|\mathbf{x}\|^2}{2}\Big). \tag{5}$$

2.4 *KeyGraph*

While KeyGraph[18] is originally proposed for keywords extraction from text data, it is extended as general co-occurrence event extraction scheme. Suppose symbolic sequence $\mathbf{D} = [e_1, \cdots, e_i][e_{i+1}, \cdots, e_j]\cdots[e_k, \cdots, e_l]$ $(i < j < k < l)$ is given, where "$[\cdot]$" is called "basket" indicating one meaningful set (e.g., one sentence). The procedure of KeyGraph consists of the following two steps:

K1 (Extracting locally influenced events). Firstly extract a pre-defined number of the most frequent events in $\mathbf{D}$ as vertices $\mathbf{V}_l$. Locally influenced graph $\mathbf{G}_l(\mathbf{V}_l, \mathbf{E}_l)$, which represents fundamental causes, can then be obtained with a pre-defined number of the most frequently co-occurring event pairs among $\mathbf{V}_l$ as edges $\mathbf{E}_l$. In the present paper, the Jaccard coefficient was used as the local co-occurrence frequency, where the counting is based on baskets.

K2 (Extracting globally influenced events). Let s be a basket, $|e|_s$ be the number of events e that appear in basket s, and $|g|_s$ be the number of events $e' \in g$

[1] The meaning of "kernel" here is a density function, while in kernel trick is a positive semidefinite Gram matrix.

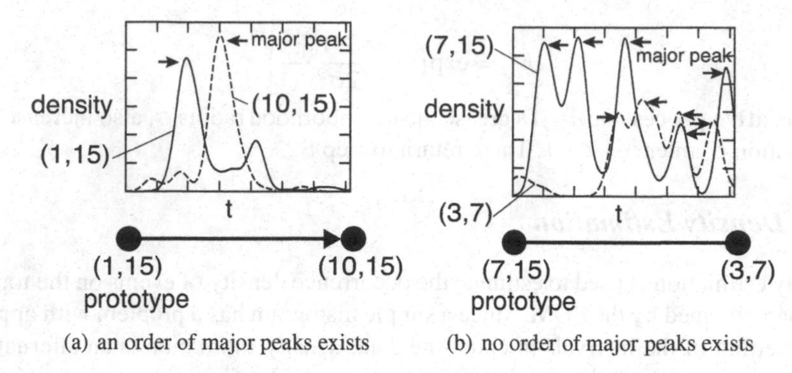

Fig. 2 A transition arrow is added between the BMUs in a KeyGraph if there exist an order of major peaks of the density distributions pair

that appear in basket s, where set of events g is composed of the connected nodes in $\mathbf{G}_l$. The conditional probability by which event $e \in \mathbf{D}$ occurred with the set of events $g \in \mathbf{V}_l$ is then defined by $global(e,g)$ as follows:

$$global(e,g) = \frac{\sum_{s \in D} |e|_s |g-e|_s}{\sum_{s \in D} \sum_{e'(\neq e) \in g} |e'|_s |e|_s}, \tag{6}$$

$$\text{where } |g-e|_s = \begin{cases} |g|_s - |e|_s & \text{if } e \in g, \\ |g|_s & \text{otherwise.} \end{cases} \tag{7}$$

The globally influenced events graph $\mathbf{G}_g(\mathbf{V}_g, \mathbf{E}_g)$ is then extracted with the predefined number of the highest $key(e)$ as $\mathbf{V}_g$, where $key(e)$ is defined by the sum of $global(e,g)$ for all clusters g, and $e \in \mathbf{V}_g$ and $e' \in g$ of the highest co-occurring event pairs are connected as $\mathbf{E}_g$. These events are not frequent but are important events in terms of the conditional probability of fundamental causes. Finally, total graph $\mathbf{G}$ is obtained by merging $\mathbf{G}_l$ and $\mathbf{G}_g$.

As an implementation of KeyGraph, Polaris[2] was used in this work.

2.5 The E^3 Algorithm

The procedure of the proposed E^3 consists of the following five steps:

E1 (Encoding). Assume an event sequence $\mathbf{D} = X_1, X_2, \cdots, X_T$ is observed. After the learning of the SOM, the best matching unit (BMU) for all events are obtained, i.e., the coordinates of the nearest prototype neuron: $(x_1, y_1), \cdots, (x_T, y_T)$.

E2 (Partitioning). Separate $\mathbf{D}$ into baskets $s = [X_t, \cdots, X_{t+l}]$. In the original KeyGraph, a basket corresponds to a set of words within a sentence. Since this procedure depends on application, this step is explained in section 3.3.2.

[2] http://www.chokkan.org/software/polaris/ (in Japanese)

E3 (Extracting co-occurrence graph). Extract a graph **G** by applying KeyGraph to **D**.

E4 (Add event transition). Estimate the occurrence density in spatio-temporal space of $(x_t, y_t, t) : t = 1, \cdots, T$, i.e., BMUs on the topology map of the SOM and time. Afterwards, add an arrow between the BMUs in a KeyGraph **G** if an order of major peaks exists in the density distribution of two nodes, i.e., prototypes (Fig. 2).

E5 (Decoding). In the graph **G**, edges and nodes that were added as transitions in the step E4 are mapped onto the topology map of the SOM by the coordinates of the prototypes.

Here, the underlying assumption is that an interpretable topology map was obtained by the SOM with density estimation. In addition, although the step E4 is currently performed manually, this is not so great a burden because KeyGraph extracts a small number of co-occurring events.

3 Application to Damage Evaluation of Fuel Cells

3.1 The Problem in Fuel Cells

The fuel cell is regarded as a highly efficient, low-pollution power generation system that produces electricity by direct chemical reaction. Solid oxide fuel cells (SOFC), in particular, have attracted a great deal of attention because they have a power generation efficiency of nearly 70% when combined with a gas turbine. However, a crucial issue in putting SOFC into practical use is the establishment of a technique for evaluating the deterioration of SOFC in the operating environment[31, 3, 13].

Since SOFC operate in harsh environments (i.e., high temperature, oxidation-reduction), the reaction area is decreased by fracture damage, and the cell performance is reduced as a result. Previously, the degree of degradation has been estimated using an electrochemical method that measures chemical degradation. Two of the co-authors have succeeded in observing mechanical damage to SOFC using the acoustic emission (AE) method[24]. Acoustic emission is an elastic wave (i.e., vibration, sound waves, including ultrasonic wave) produced by damage, such as cracks in the material, or by friction between materials. Depending on the "fracture mode" (i.e., opening or shear), the type of material, the fracture energy, the shear rate, and other factors, distinct AE wave forms are produced[15].

We previously developed the basis upon which to explore numerous AE events using the method based on a self-organizing map (SOM)[8, 6] as well as a complex network analysis[7]. In these studies, we revealed that the transition of the damage phase in SOFC also suddenly occurs as mentioned in [26]. The present study is an attempt to infer mechanical effect in SOFC from an AE event sequence. Most of the researches on AE events including other than SOFC did not focus on co-occurring AE events, but rather on typical clustering or classification tasks [5, 23, 9, 21].

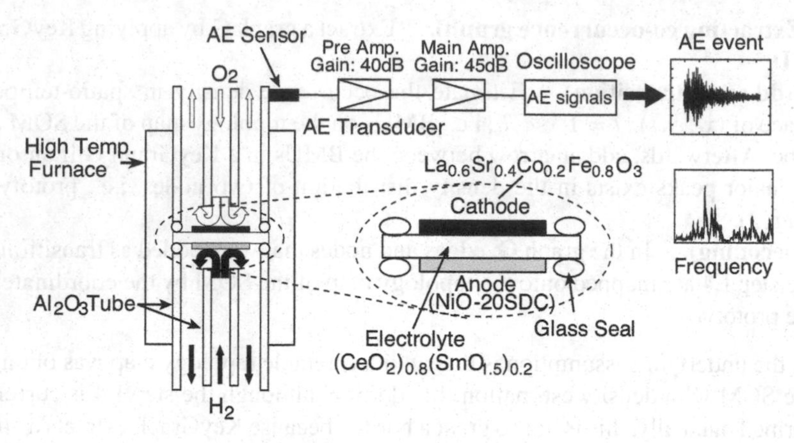

Fig. 3 SOFC damage test apparatus

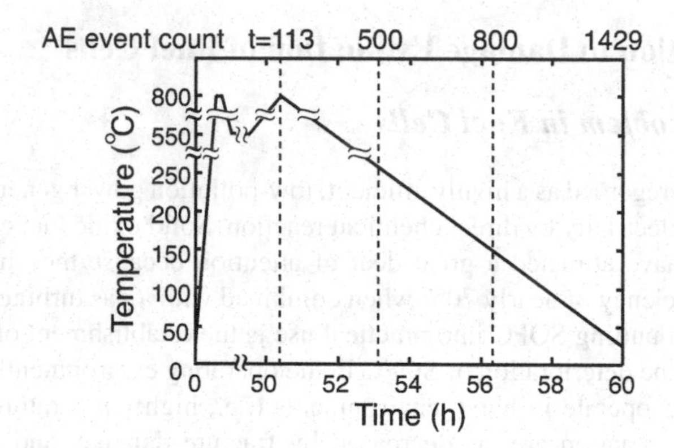

Fig. 4 Controlled temperature change and observed AE event count

3.2 Damage Evaluation Test of Fuel Cells

A schematic diagram of the apparatus used to perform the SOFC performance test is shown in Fig. 3. The test section was initially heated up to 800°C in order to melt a soda glass ring and was then gradually decreased to room temperature (Fig. 4). The AE measurement was performed using a wide-band piezoelectric transducer[3]. The AE transducer was attached to an outer Al_2O_3 tube away from the heated section. The sampling rate is 1 MHz, and so the observable maximum frequency is 500 KHz. Over 60 hours of running the SOFC, 1,429 AE events were extracted by the burst extraction method[11, 8].

Note that this damage evaluation test was to rupture the cells intentionally while lowering the temperature. Therefore, the knowledge obtained through this

[3] PAC UT-1000, URL: http://www.pacndt.com

experiment is not directly available to actual running the SOFC. However, it is sufficient to demonstrate and confirm the reasonableness of the proposed E^3.

3.3 *Adaptation of E^3 to AE Event Sequence*

3.3.1 Kernel Function

In the present work, the Kullback-Leibler (KL) divergence was used as the kernel function between frequency spectra of AE events for the kernel SOM. The KL divergence is widely known as a metric of probability distribution. We assume a frequency spectrum as a probability distribution, in the same manner that Moreno et al. applied the KL divergence to SVM for image spectrum classification[16].

Let v−discrete points of a frequency spectrum be $\mathbf{x}_i = (x_{i,1}, \cdots, x_{i,v})$. Then, the KL kernel function is defined as:

$$K_{\mathrm{KL}}(\mathbf{x}_i, \mathbf{x}_j) = \exp(-\alpha \mathrm{JS}(\mathbf{x}_i, \mathbf{x}_j)), \tag{8}$$

$$\begin{aligned} \mathrm{JS}(\mathbf{x}_i, \mathbf{x}_j) &= \mathrm{KL}(\mathbf{x}_i, \mathbf{x}_j) + \mathrm{KL}(\mathbf{x}_j, \mathbf{x}_i) \\ &= \sum_{k=1}^{v} \left\{ x_{i,k} \log \frac{x_{i,k}}{x_{j,k}} + x_{j,k} \log \frac{x_{j,k}}{x_{i,k}} \right\}, \end{aligned} \tag{9}$$

where $\mathrm{JS}(\mathbf{x}_i, \mathbf{x}_j)$ denotes the Jensen-Shannon divergence, which symmetrizes the KL divergence, and $\alpha > 0$ is a scaling parameter. Note that the spectra must be normalized as $\sum_k x_{i,k} = 1$, since KL divergence is originally for a probability distribution.

In advance, we have validated the performance of the kernel SOM with KL kernel for AE events data, using the benchmark data of damage related sounds, such as crack of a block of wood[6]. The result showed the KL kernel provides the best performance in terms of F-measure compared to the general kernel functions, such as Gaussian kernel, and the standard SOM.

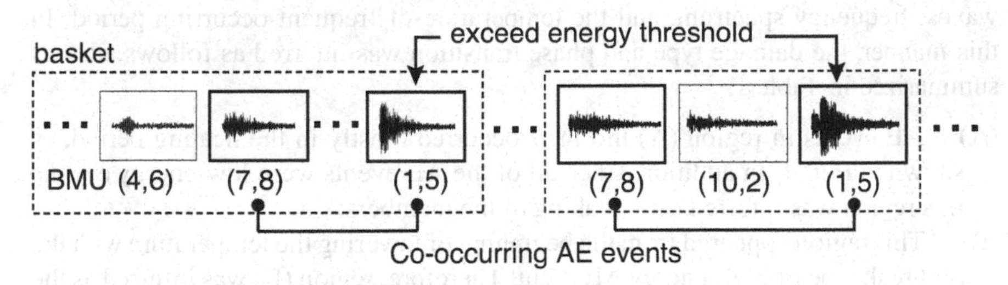

Fig. 5 Partition of an AE event sequence into baskets

3.3.2 Partition of AE Event Sequence

Assume that the potential stress in a composite material is released after a large-energy AE event occurs, i.e., interactions of internal forces are reset. In this research, the observed AE event sequence was divided into baskets followed by [19, 17], assuming a sequence until a large-energy AE event occurs to be a chain of damage progression (Fig. 5). These baskets are used in the KeyGraph.

More concretely, at first calculate the energy for all events, $E_1, E_2, \cdots, E_T$, where $E_i = \sum_j x_{i,j}^2$. Then divide the AE event sequence into baskets $s = [X_t, \cdots, X_{t+l}]$, where each basket satisfies the following condition:

$$E_{t+i} \leq E_\sigma \text{ and } E_{t+l} > E_\sigma \ (i = 0, \cdots, l-1), \tag{10}$$

where E_σ is an energy threshold. Then the AE event sequence is, for example, described as $\mathbf{D} = [\cdots(4,6),(7,8),\cdots,(1,5)]\cdots[(7,8),(10,2),(1,5)]\cdots$ in Fig. 5.

3.4 *Inference of Physical Interpretation of the Topology Map*

The number of neurons in the SOM was set to 15×15 with a regular grid. Here, the number of neurons does not affect the result if there are sufficient neurons to capture the data distribution. In addition, the parameter of the KL kernel function was set to $\alpha = 0.95$, which yields a reasonable result. Fig. 6 shows the occurrence density distribution of the AE events on the topology map of the SOM for each instant of time. The bandwidth was set to $b = 0.34$ in the KDE, which was determined by 10-fold cross validation.

The frequently occurring regions change dynamically according to time in Fig. 6. The approximate frequent occurring regions and samples of AE events are illustrated in Fig. 7. Such regions imply a certain AE type, e.g., cracking of electrolytes. Two of the co-authors, whose major fields are fuel cells and fracture mechanics, provided a physical interpretation of the topology map, by referring to actual AE waves, frequency spectrum, and the temperature of frequent occurring period. In this manner, the damage type and phase transition was inferred as follows, also as summarized in Table 1.

(A) AE events in region (A) in Fig. 7 occurred mostly in the heating period, as shown in Fig. 4. In addition, since all of the AE events were low-energy events, this region was inferred as squeaking of the members.

(B) This region appeared from the beginning of lowering the temperature with the outbreak type of high-energy AE event. Therefore, region (B) was inferred as the progression of the initial cracks, or other cracks, because of the unevenness of the materials. These AE events are may be both in the electrolyte and the electrodes.

(C) Since continuous type AE events, the frequency spectra of which are similar to the AE events in region (B), occur in this region, region (C) was inferred as squeaking of the members, followed by region (B).

(D) The AE events in this region are high-frequency events. Thus, region (D) was inferred as cracking of the electrolytes of hard materials.

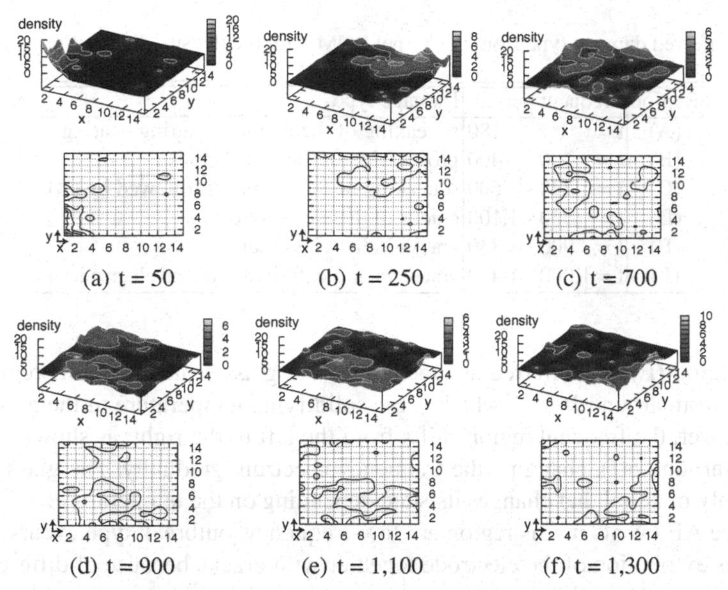

Fig. 6 Damage transition of SOFC by the kernel SOM with density estimation (upper: 3D representation, lower: 2D contour representation)

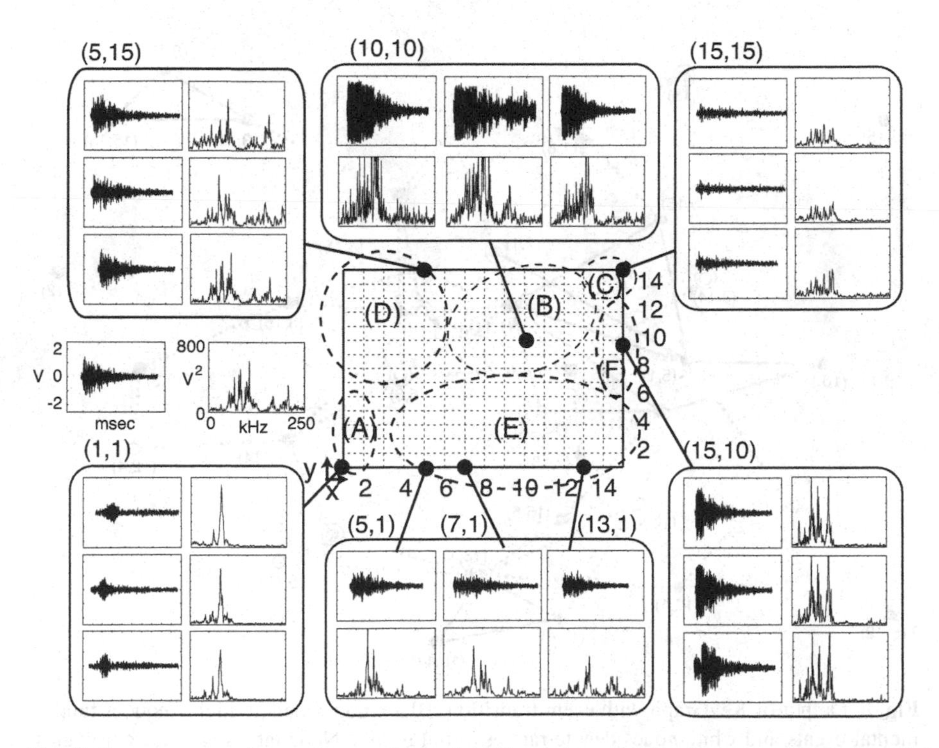

Fig. 7 Frequent occurring regions and sample AE events on the topology map by the kernel SOM

Table 1 Inferred damage type from the kernel SOM with the density distribution

Region	Frequent period	Damage type
(A)	t = 1 - 180	squeaking of the members during heating
(B)	t = 100 - 400	progression of the initial cracks
(C)	t = 220 - 600	squeaking of the members followed by (B)
(D)	t = 550 - 1,100	cracks in the electrolyte
(E)	t = 900 - 1,350	cracks in the glass seal
(F)	t = 1,000 - 1,429	cracks in and exfoliation of the electrode

(E) Region (E) was inferred as cracking of the glass seal. Since this region appears from around 100°C, which is the solidifying temperature of the glass seal. Moreover, the frequent region shifts from the left to the right, as shown in Figs. 6(d) through 6(f), changing the frequency spectrum gradually. The glass seal is the only material that changes its state depending on the temperature.

(F) The AE events in this region are low-frequency, outbreak type events, which means exfoliation of the electrode together with cracks because of difference of heat contraction ratios.

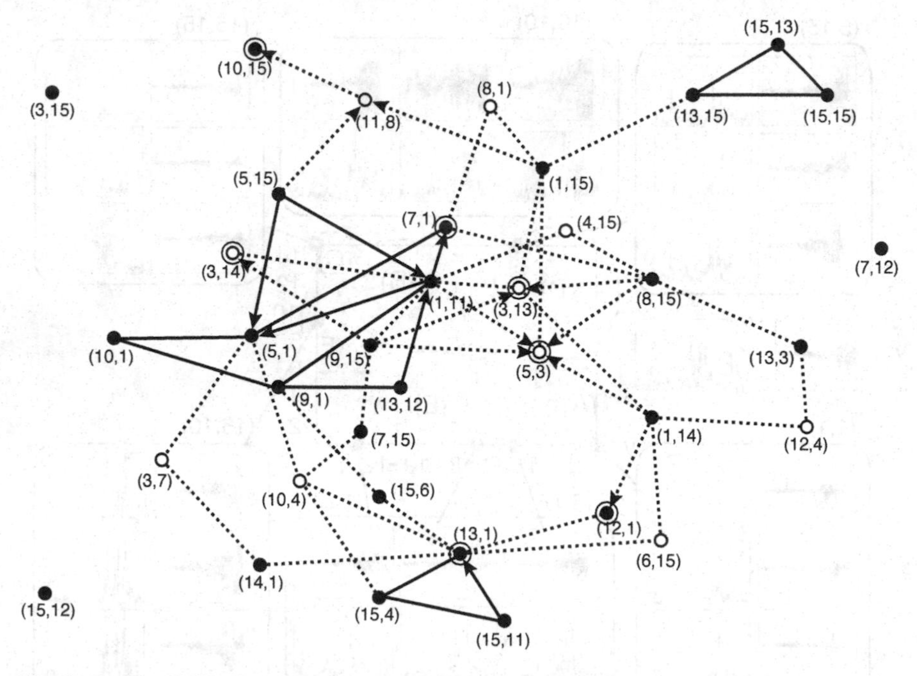

Fig. 8 Output of KeyGraph with event transition. Black nodes denote high frequent fundamental events and white nodes denote rare essential events. Node labels are the coordinates of prototypes by the kernel SOM

3.5 Mechanical Effects Inferred by E^3 Analysis

The output of KeyGraph is shown in Fig. 8. The energy threshold E_σ was set to 1,500 from among 1,000, 1,250, 1,500, 1,750, and 2,000, the effect of the energy threshold is discussed in section 3.6. A black node indicates a prototype of a locally influenced event, i.e., fundamental event, whereas a white node indicates a prototype of a globally influenced event. A solid line indicates higher occurrence frequency than a dotted line. Event transitions, which are shown by arrows in the graph, were then added by referring to the density change of the prototype events pairs (Fig. 9). An ending event of the transitions is denoted as a double circle.

These extracted event transitions were mapped onto the topology map shown in Fig. 10. Fig. 10(a) shows the transitions from region (B) to region (D), indicating that the progression of the initial cracks affects the electrolyte, while Fig. 10(b) shows the transitions from region (D) to region (B) in the direction opposite that shown in Fig. 10(a). Thus, mutual interaction exists in the progression of the initial cracks and the electrolyte. Fig. 10(c) shows transitions from region (B) to region (E) via (D), indicating that the progression of initial cracks affects the electrolyte same as in Fig. 10(a) and in addition cracks of the electrolyte affects the glass seal. Moreover, region (E) which is cracks of the glass seal is affected by electrolyte and electrode whose regions are (D) and (F) as illustrated in Fig. 10(c) to Fig. 10(f). On the other hand, there is no influence from regions (A) and (C) on any other region. This is reasonable because these regions are inferred as squeaking of the members. In addition, the glass seal does not affect the other materials, and only the glass seal is affected by the other materials. Moreover, a result that is interesting

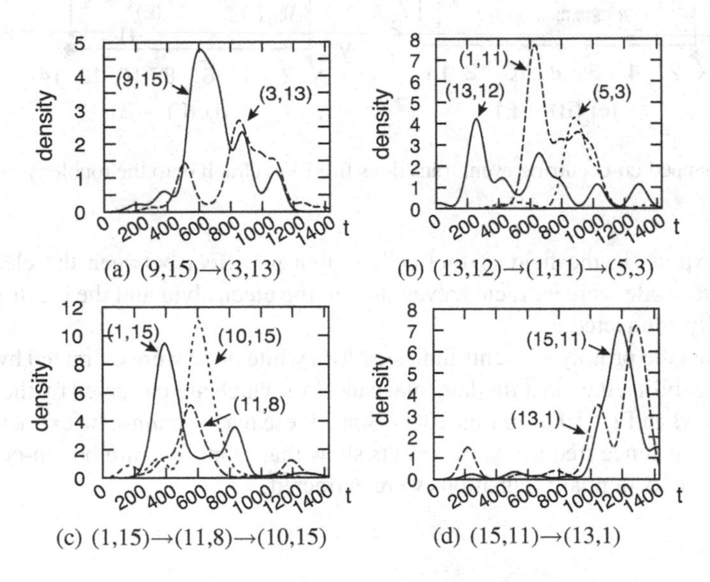

Fig. 9 An order of major peaks on occurrence density change of the prototypes

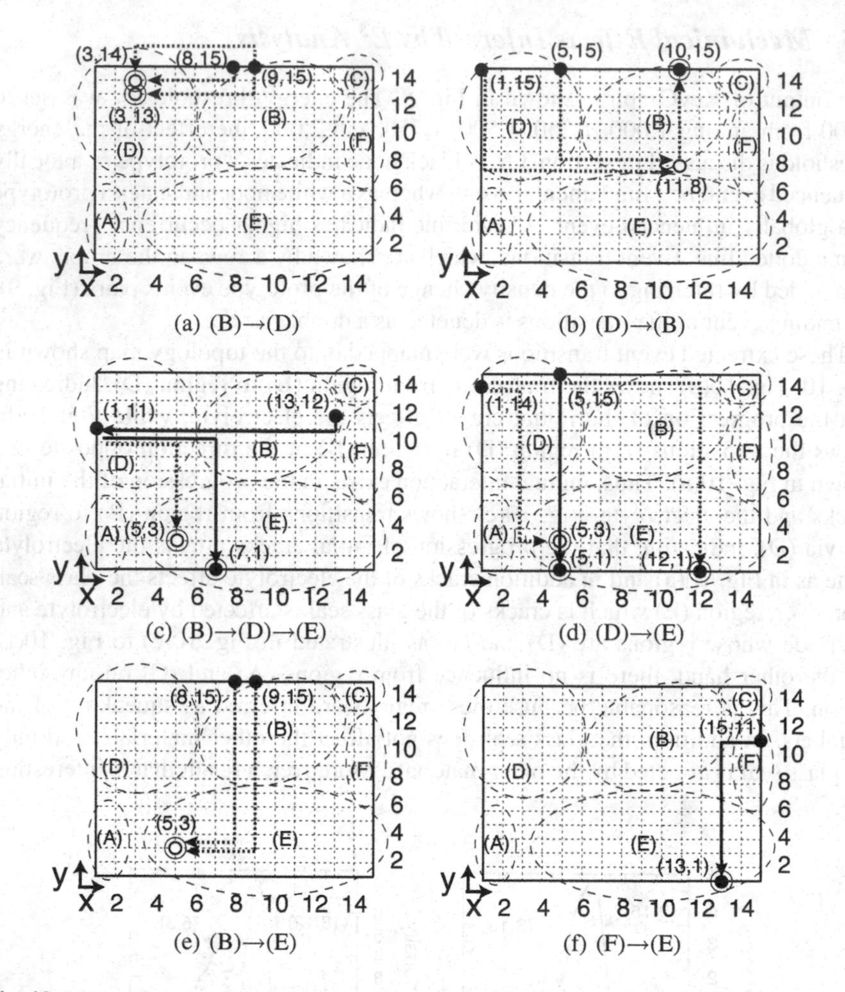

Fig. 10 Mapped co-occurring event transitions from KeyGraph onto the topology map

to even experts in the field of fuel cells is that no effect between the electrolyte and the electrode were extracted, even though the electrolyte and the electrodes are structurally connected.

Note that the prototype events indicated by a white node were extracted by conditional probability, the most fundamental causes of which are indicated by the origins of the arrows in Fig. 10(a) to Fig. 10(e). Since these events cannot be extracted only by their occurrence frequency, the results show that relatively rare but co-occurring essential events in phase transitions were extracted.

Table 2 Occurrence/co-occurring frequency of transition events

event transition $a \rightarrow b$	$N(a), N(b)$	$N(a \cap b)$
Co-occurring LI events		
(1,11)→(7, 1)	9, 14	4
(5,15)→(5, 1)	11, 11	4
(1,14)→(12, 1)	9, 15	4
(13,12)→(1,11)	9, 9	4
(15,11)→(13, 1)	9, 11	4
Co-occurring GI events		
(9,15)→(3,14)*	9, 5	4
(1,11)→(5, 3)*	9, 5	4
(1,15)→(11, 8)*	9, 8	3
(5,15)→(11, 8)*	11, 8	3
(11, 8)*→(10,15)	8, 11	3
(8,15)→(3,13)*	9, 7	2
(9,15)→(3,13)*	9, 7	2
(1,14)→(5, 3)*	9, 5	2
(8,15)→(5, 3)*	9, 5	2
(9,15)→(5, 3)*	9, 5	2

Table 3 Parameter settings of the KeyGraph

parameter \ E_σ	1,000	1,250	1,500	1,750	2,000
LI events	30	28	26	26	24
LI event pairs	32	35	26	25	26
GI events	10	10	10	10	10
GI event pairs	27	28	29	24	23

Table 4 Effect of the energy threshold

transition \ E_σ	1,000	1,250	1,500	1,750	2,000
(B) → (D)	3	8	4	4	4
(D) → (B)	3	1	2	1	0
(B) → (E)	1	1	2	0	0
(D) → (E)	4	3	5	4	5
(E) → (D)	1	0	0	0	1
(F) → (E)	1	1	1	1	1
total	13	14	14	10	11

3.6 Extracted Essential Rare Events

Table 2 shows the occurrence frequency of the extracted transition events from the above experiment. The listed event transitions correspond to Fig. 10. Here, $N(a)$ denotes the occurrence frequency of event 'a', where the frequency means the number of baskets containing event 'a' at least one. Locally and globally influenced events, which are denoted by LI and GI, were extracted respectively by steps K1 and K2 of the KeyGraph algorithm. GI events are marked as a^* in table 2.

While five transitions of co-occurring LI events were extracted, that of co-occurring GI events was ten. This fact means that co-occurring AE events during damage phase transitions were extracted more by conditional probability of fundamental AE events rather than by high frequent fundamental events.

3.7 Effect of Energy Threshold

This section discusses the effect of energy threshold E_σ which is a parameter to divide an AE event sequence into baskets. The larger E_σ is the more events obtain in one basket and the less number of basket.

The parameter settings of the KeyGarph for every E_σ is listed in Table 3. The occurrence frequency of AE events was fixed by nine for every E_σ, because the number of prototypes which has more than ten AE events as BMU was relatively few, while more than eight was too many with this SOM learning result. From this fact, the number of LI events were 24 to 30. The number of co-occurring LI event pairs, which are links in the KeyGraph, were set to around the number of LI events. Also GI events was set to 10, and co-occurring GI event pairs were set to around the number of LI events.

With these settings, in order to investigate the effect of E_σ to the inferred mechanical effects, we compared the number of event transitions extracted by E^3 as shown in Table 4. The appropriate E_σ is around 1,250 to 1,500 as transition (B)→(E) disappears above 1,750 and (E)→(D) appears 1,000 and 2,000. (E)→(D) suppose to appear by chance depending on relative frequency when E_σ is an inappropriate setting. Whereas, (F)→(E) appeared not by chance since it appears in every E_σ. E_σ =1,500 is well balanced in terms of the number of appeared transitions. Also

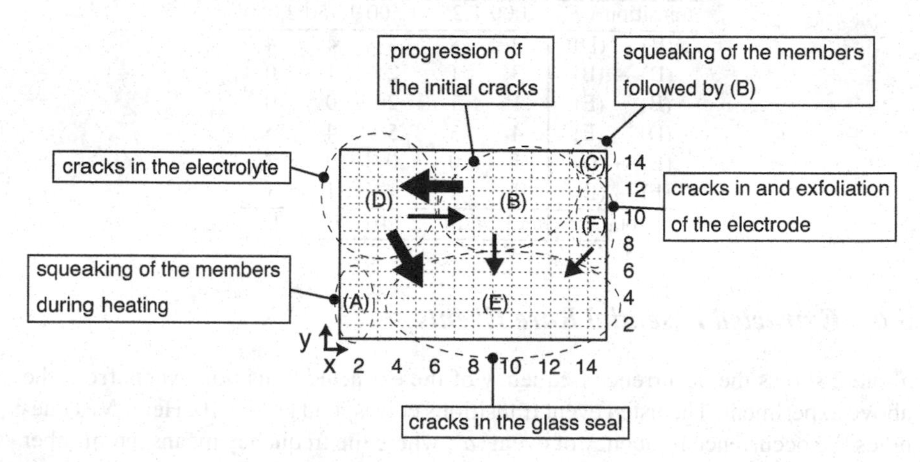

Fig. 11 Inferred mechanical effects in SOFC by E^3 analysis

the important fact is that E_σ is not sensitive to the final inferred mechanical effects, even though the output of KeyGraph is different.

Considering above discussion, the final inferred mechanical effects is illustrated in Fig. 11. The thick arrow in the map indicates strong effect and thin arrow indicates weak effect.

3.8 Scenario

This section describes a scenario of the damage process of SOFC inferred from our E^3 analysis (Fig. 11 together with Table 1).

1) Stable running period: Although squeaking of the members of the materials occur (region (A)), these events do not affect damage of the cells at all. That is, these events can be disregarded as noise events when monitoring the running.
2) Primary stage of lowering the temperature: At the beginning of lowering the temperature to stop running, it begins to progress the initial small cracks because of unevenness of the material (region (B)). The attendant squeaking of the members, region (C), do not affect any other damage as well.
3) Secondary stage: With further lowering the temperature, stress is accumulated in the heat-shrinkable electrolyte. Then, the cracks in the electrolyte are triggered by the progression of the initial cracks ((B)→(D)). In contrast, though cracks of the electrolyte promote the initial cracks, its effect is lesser ((D)→(B)).
4) Latter stage: As solidifying the glass seal, the glass seal accumulates stress strained by the electrolyte. Then the glass seal is damaged by releasing the accumulated internal stress with the trigger of cracks of the electrolyte ((D)→(E)).
5) Final stage: Although exfoliation together with cracks develop in the electrode (region (F)), these are not affected from cracks of the electrolyte. Also progression of the initial cracks and cracks of the electrode promote the damage of the glass seal ((B)→(E) and (F)→(E)).

4 Future Perspective

We demonstrated the proposed E^3 using an AE event sequence observed by the damage evaluation test on SOFC. The kernel SOM with density estimation provides a comprehensive topology map that can infer the damage type and damage phase transitions within the map. Afterwards, the E^3 extracted and suggested AE events on the map that are not frequent themselves, but co-occurred in phase transitions. These AE events cannot be extracted only by the method based on the occurrence frequency of a single event, e.g., density estimation.

The domain experts can reasonably explain the results as mechanical effects and ascertain novel information. It is great advance that the domain experts can form a hypothesis via E^3. This hypothesis can be verified by checking the reproducibility of the event sequences via several damage tests.

In the proposed E^3, event transitions are manually assigned by referring to the occurrence density change of prototypes. In the future, activity propagation mechanisms, such as priming activation indexing (PAI)[14], are an option for the natural introduction of event transition to E^3.

The proposed E^3 can contribute to the clarification of the fracture mechanism or to monitoring the phase transition point or fatal damage. In the future, the fracture mechanism in the SOFC will be clarified by comparing several experimental conditions and/or by combining computational simulation.

5 Conclusion

We proposed the essential event extractor (E^3) scheme for a non-symbolic event sequence to extract relatively rare but co-occurring events in phase transitions that exhibit hidden forces. The self-organizing map (SOM) is used as vector quantization (VQ) to encode non-symbolic events and KeyGraph as a co-occurrence graph. Together with density estimation on the topology map of the SOM, co-occurring event sequences can be obtained on the map. The E^3 enhances the co-occurring analysis of a symbolic sequence to a non-symbolic sequence, because E^3 requires only the dissimilarity between events.

We demonstrated E^3 by applying to an acoustic emission (AE) event sequence observed from damage test of fuel cells. Consequently, mechanical effects in the fuel cells can be inferred by the result of E^3 analysis, and these extracted effects express hidden forces that appear during the damage phase transitions.

Acknowledgment

This work was supported in part by the Management Expenses Grants for National Universities Corporations and also by KAKENHI (21700165) both from the Ministry of Education, Culture, Sports, Science and Technology of Japan (MEXT).

References

1. Agarwal, D., Broder, A., Chakrabarti, D., Diklic, D., Josifovski, V., Sayyadian, M.: Estimating rates of rare events at multiple resolutions. In: Proc. of the 13th ACM SIGKDD International Conference on Knowledge Discovery and Data Mining (KDD 2007), pp. 16–25 (2007)
2. Agrawal, R., Srikant, R.: Fast algorithms for mining association rules. In: Proc. of the 20th International Conference on Very Large Databases (ICVD 1994), pp. 487–499 (1994)
3. Atkinson, R., Ramos, T.M.G.M.: Chemically-induced stresses in ceramic oxygen ion-conducting membranes. Journal of Solid Sate Ionics **129**, 259–269 (2000)
4. Boulet, R., Jouve, B., Rossi, F., Villa, N.: Batch kernel SOM and related Laplacian methods for social network analysis. Neurocomputing 71, 1257–1273 (2008)

5. Emamian, V., Kaveh, M., Tewfik, A.H., Shi, Z., Jacobs, L.J., Jarzynski, J.: Robust clustering of acoustic emission signals using neural networks and signal subspace projections. Journal on Applied Signal Processing 2003(3), 276–286 (2003)
6. Fukui, K., Sato, K., Mizusaki, J., Numao, M.: Kullback-leibler divergence based kernel som for visualization of damage process on fuel cells. In: Proc. of 22th IEEE International Conference on Tools with Artificial Intelligence, ICTAI 2010 (2010)
7. Fukui, K., Sato, K., Mizusaki, J., Saito, K., Kimura, M., Numao, M.: Growth analysis of neighbor network for evaluation of damage progress. In: Theeramunkong, T., Kijsirikul, B., Cercone, N., Ho, T.-B. (eds.) PAKDD 2009. LNCS (LNAI), vol. 5476, pp. 933–940. Springer, Heidelberg (2009)
8. Fukui, K., Sato, K., Mizusaki, J., Saito, K., Numao, M.: Combining burst extraction method and sequence-based som for evaluation of fracture dynamics in solid oxide fuel cell. In: Proc. of 19th IEEE International Conference on Tools with Artificial Intelligence (ICTAI 2007), pp. 193–196 (2007)
9. Godin, N., Huguet, S., Gaertner, R.: Influence of hydrolytic ageing on the acoustic emission signatures of damage mechanisms occurring during tensile tests on a polyester composite: Application of a kohonen's map. Composite Structures 72(1), 79–85 (2006)
10. Han, J., Kamber, M., Pei, J.: Data Mining, Concepts and Techniques, 2nd edn. Morgan Kaufmann, San Francisco (2006)
11. Kleinberg, J.: Bursty and hierarchical structure in streams. In: Proc. the 8th ACM SIGKDD International Conference on Knowledge Discovery and Data Mining (KDD 2002), pp. 1–25 (2002)
12. Kohonen, T.: Self-Organizing Maps. Springer, Heidelberg (1995)
13. Krishnamurthy, R., Sheldon, B.W.: Stresses due to oxygen potential gradients in nonstoichiometric oxides. Journal of Acta Materialia 52, 1807–1822 (2004)
14. Matsumura, N., Ohsawa, Y., Ishizuka, M.: Pai: Automatic indexing for extracting asserted keywords from a document. New Generation Computing 21, 37–47 (2003)
15. Miller, R.K., Kill, E.V.K., Moore, P.O., Hill, E.V.: Acoustic Emmision Testing. American Society for Nondestructive (2005)
16. Moreno, P.J., Ho, P.P., Vasconcelos, N.: A kullback-leibler divergence based kernel for svm classification in multimedia applications. Advances in Neural Information Processing Systems 16 (2003)
17. Ohsawa, Y.: Keygraph as risk explorer from earthquake sequence. Journal of Contingencies and Crisis Management 10(3), 119–128 (2002)
18. Ohsawa, Y., Benson, N.E., Yachida, M.: Keygraph: Automatic indexing by cooccurrence graph based on building construction metaphor. In: Proc. Advances in Digital Libraries Conference, pp. 12–18 (1998)
19. Ohsawa, Y., Yachida, M.: Discover risky active faults by indexing an earthquake sequence. In: Arikawa, S., Nakata, I. (eds.) DS 1999. LNCS (LNAI), vol. 1721, pp. 208–219. Springer, Heidelberg (1999)
20. Oja, M., Kaski, S., Kohonen, T.: Bibliography of self-organizing map (som) papers: 1998-2001 addendum. Neural Computing Surveys 3, 1–156 (2002)
21. Omkar, S., Karanth, R.: Rule extraction for classification of acoustic emission signals using ant colony optimisation. Engineering Applications of Artificial Intelligence 21, 1381–1388 (2008)
22. Phan, X.H., Nguyen, L.M., Ho, T.B., Horiguchi, S.: Improving discriminative sequential learning with rare-but-important associations. In: Proc. of the 11th ACM SIGKDD International Conference on Knowledge Discovery and Data Mining (KDD 2005), pp. 304–313 (2005)

23. Rippengill, S., Worden, K., Holford, K.M., Pullin, R.: Automatic classification of acoustic emission patterns. Journal for Experimental Mechanics: Strain 39(1), 31–41 (2003)
24. Sato, K., Omura, H., Hashida, T., Yashiro, K., Kawada, T., Mizusaki, J., Yugami, H.: Tracking the onset of damage mechanism in ceria-based solid oxide fuel cells under simulated operating conditions. Journal of Testing and Evaluation 34(3), 246–250 (2006)
25. Silverman, B.: Density Estimation for Statistics and Data Analysis. Chapman & Hall, Boca Raton (1986)
26. Sornette, D.: Predictability of catastrophic events: Material rupture, earthquakes, turbulence, financial crashes, and human birth. Proc. the National Academy of Sciences of the United States of America (PNAS) 99(Suppl. 1), 2522–2529 (2002)
27. Tukey, J.W.: Exploratory Data Analysis. Addison-Wesley, Reading (1977)
28. Vilalta, R., Ma, S.: Predicting rare events in temporal domains. In: Proc. of the 2002 IEEE International Conference on Data Mining (ICDM 2002), pp. 474–481 (2002)
29. Weiss, G.M., Hirsh, H.: Learning to predict rare events in event sequences. In: Proc. of the 4th ACM SIGKDD International Conference on Knowledge Discovery and Data Mining (KDD 1998), pp. 359–363 (1998)
30. Xu, R., Wunsch, D.C. (eds.): Clustering. SCI. IEEE Press, Los Alamitos (2008)
31. Yasuda, I., Hishinuma, M.: Lattice expansion of acceptor-doped lanthanum chromites under high-temperature reducing atmospheres. Electrochemistry 68(6), 526–530 (2000)

Detecting Car Accidents Based on Traffic Flow Measurements Using Machine Learning Techniques

L.D. Tavares, G.R.L. Silva, D.A.G. Vieira,
R.R. Saldanha, and W.M. Caminhas

Abstract. This paper deals with the problem of detecting the occurrence of a car accident in an urban environment. Firstly, a model based on Cellular Automata is designed to simulate the traffic flow with its main features such as: multiple lanes, cars, traffic lights, buses and bus stops. Afterwards, machine learning techniques are trained with the traffic flow measurements considering both the normal and the situation in which the accident caused a partial closure of the lanes. Several machine learning techniques results are presented to several car breaking scenarios.

1 Introduction

The land transportation system is an important resource for the country economy and population well-being, thus, when this system does not work well, several sectors are affected. Considering, for instance, the urban transit system of a Brazilian large city, such as São Paulo or Belo Horizonte, this problem can be even more serious. In these cities the most common problem is related to congestion. Congestion can be generated when the number of vehicles is greater than the capacity of the road or for any momentary interruption (accidents or maintenance of the road). Therefore, it is necessary to develop tools that can detect the moment and place these problems occur. Hence, a corrective action can be taken in order to returns the flow to its normal state. The objective of this paper is to conduct a comparative study of different classifiers in order to detect congestion in an urban traffic.

L.D. Tavares · G.R.L. Silva · D.A.G. Vieira · R.R. Saldanha · W.M. Caminhas
Dep. of Electrical Engineering, Universidade Federal de Minas Gerais, Brazil
e-mail: tavares@cpdee.ufmg.br

D.A.G. Vieira
ENACOM - Handcrafted Technologies, Brazil
e-mail: douglas.vieira@enacom.com.br

I. Hatzilygeroudis and J. Prentzas (Eds.): Comb. of Intell. Methods and Appl., SIST 8, pp. 109–124.
springerlink.com

For this study it was built a simulator of Urban Traffic flow using Cellular Automata (CA), called Cellular Automata for Urban Traffic Simulation (CAUTS). This model considers the presence of cars, trucks, traffic lights, buses and bus stops.

CA is, in short, the mathematical model discrete in time, space and states. Its fundamental unit is called *cell*. This kind of model is based on two simple components: local rules and neighborhood. Local rules are responsible for calculating the next state of the cell, based on the influence of its neighborhood. Only with those components CA can reproduce (simulate) dynamic complex systems, ranging from biology to chemical reactions [1]. CAUTS has resources capable of simulating most of the features of an urban traffic as main roads, secondary roads, traffic lights and bus stop. Moreover, it is possible to generate events that cause traffic jams, such as stopped vehicles and accidents, which is the main focus of this work. The database was tested with different methods of classification, so that it can detect which part of the model and at what time an incident occurred. The classifiers used were: (i) Naïve Bayes (NB), (ii) Decision-Tree (DT), (iii) K-Nearest Neighbor (K-NN), (iv) Multilayer Perceptrons (MLPs), (v) Support Vector Machine (SVM), (vi) Adaptive Neuro-Fuzzy Inference Systems (ANFIS). The paper is organized as follows. Sections 4 and 5 show the basic concepts employed in the construction of CAUTS model. Then, Section 7 contains the results obtained, considering several different scenarios. Finally, the conclusion and future works are in Section 8.

2 Overview on the Traffic Flow Theory

Theories of Traffic Flow seek to study and describe the relationships between the vehicles, routes, components and infrastructure as traffic lights, signs, among others, in mathematicians terms. These theories emerged in the 30's in an attempt to relate the magnitudes of flow density and velocity, by scientist Bruce Greenshields. Today these theories are based all the tools and models of traffic flow [2]. The applications of these theories are broad. Among them are:

- Evaluation of alternative treatments in traffic management;
- Design and testing of new lanes;
- Models operational flow serving as a sub-module in other tools (model-based traffic control and optimization and dynamic traffic assignment);
- Traffic management training.

Papageorgiou [3] explains the phenomena that do not always observed in traffic are evident to the correct equation, and still divides approaches into three categories:

1. Purely deductive approaches: where it is necessary to know the laws of physics exist that govern the phenomenon;
2. Purely inductive approaches: where real systems input/output pairs data are available, which are then used to adjust general mathematical structures as ARIMA models, neural networks and polynomial approximations, for example;
3. Intermediate approaches: where the first structural model exists and then real data are used to adjust the model.

Whatever the approach taken, it is still possible to classify it using the following criteria [4]:

- Type of variables;
- Level of Detail;
- Representation of the process;
- Operationalization;
- Range of application.

Type of variables is defined on the bais of how it treats the passage of time, ie environmental change occurs in a continuous or discrete. The *level of detail* with respect to how the approach works existing entities in the model. The microscopic approaches have a greater level of detail, where the entities are treated individually. The macroscopic approaches have little level of detail and understand the traffic as a whole. In an intermediate level of detail exist mesoscopic approaches where blocks entities are treated as platoons. The *representation of the process* is characterized by existence of random variables. When they are not present the process is deterministic, on the other hand, if they are needed the process is stochastic.

The criterion of operation checks whether the approach is analytical or by simulation. Finally, the scale defines the scope of application of the approach, for example, a city, an avenue, or simply a stretch of street.

3 Cellular Automata

Studies on the potential of CA started around 1950's by von Neumann and Ulam [1]. CA is, in short, a mathematical model discrete in time, space and states. Its fundamental unit is called *cell*. This kind of model is based on two simple components: local rules and neighborhood. Local rules are responsible for calculating the next state of the cell, based on the influence of its neighborhood. Only with those components CA can reproduce (simulate) dynamic complex systems, ranging from biology to chemical reactions [1].

The simplest case of elementary CA is an one-dimensional array of cells, where each cell can have the values 0 or 1. Consider a_i^t as the state of the cell of index i at the moment t, an example of local rule δ for this elementary CA is [1] [5]:

$$a_i^{t+1} = \delta\left(a_{i-1}^t, a_i^t, a_{i+1}^t\right) \quad (1)$$

Formally, CA is defined by a tuple $A = (S, d, n, \delta)$, where S is set of states that one cell can assume, d is the dimension, n is the influence of the neighborhood structure over a cell a, and δ is the local rule. In a uniform CA model the same δ function is applied over all cells, but it is possible to have different local rules for distinct sets of cells, in this case has a non-uniform or hybrid CA [6]. CA may also include stochastic elements, such as probabilistic local rules, as shown below:

$$\delta(.) = \begin{cases} s_1, \, if \; p \\ s_2, \, if \; 1 - p \end{cases} \quad (2)$$

where p is the probability of occurrence of the state $s_1 \in S$.The update of the cells may occur in a synchronous or asynchronous form. The CA is classified as synchronous if all cells are updated at the same time, but, if some parts of the model are updated at different times the CA is classified as asynchronous.

The best known application, based on CA, is the "Game of Life", created in 1970 by Conway [1] [7]. In this game each cell is a unicellular organism that can assume one of two states: 0 - dead or 1 - alive. The local rules of this game are:

- (R1) Under population: any living cell will die if it has less than two alive neighbors;
- (R2) Overcrowding: any living cell will die if it has more than three alive neighbors;
- (R3) Perpetuation: any cell will remain for the next generation if it has two or three neighbors;
- (R4) Reborning: any dead cell will revive if it has exactly three live neighbors.

Other examples of applications based on CA are well documented in the literature, see, for instance, [8, 9, 10].

4 Simulator Features

Among several methods to traffic flow simulations, the ones based on the use of Cellular Automata (CA) have received an especial attention of researchers. Some papers from the 1990's presented the bases concerning the use of CA for traffic flow [11, 12, 13, 14, 15]. These results considered the basic acceleration, deceleration, velocity randomization and velocity update rules. A review considering road traffic flow can be found in[16]. It shows that most of the concerns are related to acceleration, deceleration and lane changes for freeways. Makowiec and Miklaszewski [17] added supplementary rules to the traditional model such a way to increase the mean velocity. It is expected that most of drivers want to travel as close as possible to the maximum allowed speed. The CA is a very useful and efficient method, and can be applied to online simulation of traffic flow, as presented in [18]. In [19] it was derived the critical behavior of a CA traffic flow model by means of an order parameter breaking the symmetry of the jam-free phase. Fuks [20] considered a deterministic CA model and derived a rigorous flow at arbitrary time. Other important aspect is the jamming caused by the reduction of the number of lanes. This reduction can be due to repairing, accidents and even because it is part of the road design. Studying the road capacity, Nassab et. al. [21] considered a road partial reduction from two lanes to one lane. The blockage of one lane, caused by an accident car, was recently studied in [22]. This paper considers the study of a car accident in an urban environment. By urban environment it is required to consider: (i) multi-lane traffic flow; (ii) crossroads; (iii) traffic lights; (iv) trucks; (v) buses and; (vi) bus stops. The presence of buses and bus stops requires specific rules. These rules are important to the traffic flow in urban areas.

5 Model Definition

The model of urban traffic flow is implemented based on a two-dimensional Stochastic Cellular Automata, called Cellular Automata for Urban Traffic Simulation - CAUTS. CAUTS has resources capable of simulating the features of an urban traffic as main roads, secondary roads, traffic lights and bus stop. Moreover, it is possible to generate events that cause traffic jams, such as stopped vehicles and accidents. The sub sections below will detail the proposed model.

5.1 Maps Definitions

The cell of the model can represent one of two states: 0 - empty, 1 - occupied. All cells of the model are square with side equal to 5.5 meters. This measure represents the average sized car in the Brazil, taking into account the distance between cars. The properties of cells are defined as a triple: $c_{i,j} = \{pd, sd, vmax\}$, where: (i) pd is the predominant direction; (ii)sd is the secondary direction; (iii) $vmax$ is the speed limit. For predominant direction, means, the direction in which the vehicle will stay longer; and, by secondary direction the change route or direction, such as lane-changing or street change. The speed limit determines how many cells can be advanced forward, at most, per iteration. Each direction d has a code, and their respective shift in the axis x and y, as can be illustrated in the Figure 1. Moreover, it allows a vehicle to move forward up to 3 cells. To indicate that a cell is not available for transit and the end of road (cell where vehicle is removed from model), two triples, $\{0, 0, 0\}$ and $\{9, 9, 9\}$, are used, respectively.

5.2 Environmental Rules

These are the rules that change a set of cells to implement some desired characteristics. One of the most important feature in urban traffic is the presence of traffic lights. Consider the complementary set of traffic lights T_1 and T_2, where the cells affected by these sets are defined as $T_1 = \{(x_1, y_1), (x_2, y_2), \ldots, (x_n, y_n)\}$. Similarly, consider T_2, where, for example, T_1 is the set of traffic lights in the main road and T_2 in the secondary road. The complementarity, then, is defined by: $T_1(Green) \Rightarrow T_2(Red)$, $T_1(Yellow) \Rightarrow T_2(Red)$, $T_2(Green) \Rightarrow T_1(Red)$, $T_2(Yellow) \Rightarrow T_1(Red)$. The Equation 3 shows how the traffic lights can be modeled.

$$(RT): \begin{cases} T(Red) & \Rightarrow vmax_T = 0 \\ T(Yellow) & \Rightarrow vmax_T = 1 \\ T(Green) & \Rightarrow vmax_T = vmax \end{cases} \tag{3}$$

$\forall (x, y) \in T$.

As mentioned above, the model contains features that considers broken vehicles or accidents. Consider the set $A = \{(x_1, y_1), (x_2, y_2), \ldots, (x_n, y_n)\}$ as the location of the cells where the incident occurs, at time t_0 with k iterations long. Additionally, consider $cod(A, t_{0-1}) = cod(A, t_{0-1})$ as the cell triple code before the incident. Since the cells not available for transit is represented using $\{0, 0, 0\}$, then, the presence of stopped vehicle is modeled as:

$$(RA): \begin{cases} cod_{(A,t)} = 000, & if\ t \leq t_0 + k \\ cod_{(A,t)} = cod_{(A,t_0-1)}, & otherwise \end{cases} . \tag{4}$$

At the beginning and the end of each road there is one sensor. These sensors are responsible for capturing the statistics, such as, number of vehicles (flow) and their speeds.

5.3 Vehicles Definitions

The model implemented has three types of vehicles: small vehicles, like cars, and large vehicles, such as buses and trucks. Small vehicles occupy only one cell, while large vehicles occupy three cells in length and the width of one cell. Currently, the model considers that large vehicles can only move in the main roads and can not switch lanes or routes. Buses and trucks differ, themselves, by the fact that buses have to stop at bus stops. The vehicle models have the following structure: (i) kind of vehicle: $1 - car$, $2 - bus$ *or* $3 - truck$; (ii) vehicle location (x, y); (iii) lane change indicator (t_1); (iv) vehicle current speed (vel_i); (v) time of the vehicle last stopped (t_2); (vi) sensor identifier (sid). The feature (iii) is applied only when the vehicle is a car and indicates how many iterations has passed from the time vehicle last changed a lane. This serves to prevent the car change its lane by consecutive times. Because of this, the model does not allow a vehicle leaving right lane and go to left

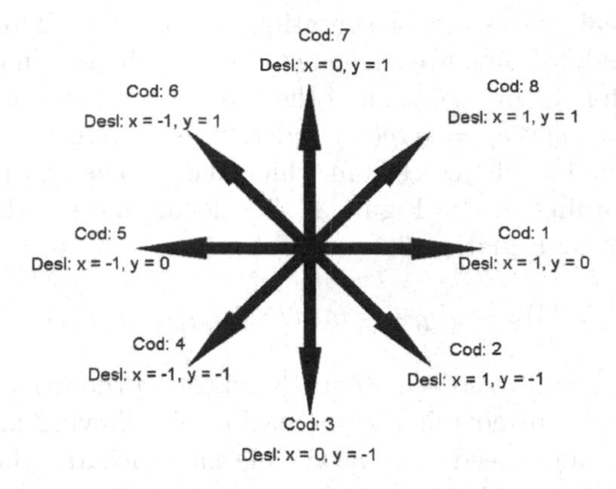

Fig. 1 Code and respective dislocation (in x and y axis).

lane, whereas there is a central lane, instantly. Consider a vehicle v_i in the set $V = \{v_1, v_2, \ldots v_i, \ldots v_n\}$ at the moment t. The location of the vehicle may be recovered by the expression:

$$loc_i = \begin{cases} [x_1, y_1] = posi(v_i), & if\ v_i = car \\ \begin{bmatrix} x_1\ y_1 \\ x_2\ y_2 \\ x_3\ y_3 \end{bmatrix} = posi(v_i), \ otherwise \end{cases} . \tag{5}$$

Consider the location of all vehicles as $LOC = posi(V)$. The function $dir_i = direc(v_i)$, where $dir_i = [xd_1, yd_1]$ for small cars, indicates the vehicle moving, according to the Figure 1. For instance, a vehicle is moving to the east, the function $direc(.)$ will be $[xd, yd] = [1, 0]$ and $[xd, yd] = [-1, 1]$ for northwest moving. The current speed of the vehicle vel_i is accessed through the function $speed(v_i)$. The maximum speed that a vehicle can achieve depends on its type and its location at time t, as small cars tend to be faster than large vehicles in urban traffic. The speed is computed as cells/iteration, of c/i. The speed limit is calculated by the function $vmax_i = velocmax(v_i, loc_i)$. The Equation 6 defines the rule for local acceleration. This rule represents the intention of the driver to speed up as much as possible, i.e., the speed limit of the road will be respected.

$$(R1) : vel_{(i,R1)} = min\left(vel_{(i,t)} + 1, vmax_i\right) \tag{6}$$

However, we know that drivers may, so seemingly random, reduce vehicle speed. Consider α as the probability of a slowing down, then the local rule for this event is given by 7.

$$\begin{array}{l} if\ rand < \alpha_i, \\ (R2) : \ vel_{(i,R2)} = max\left(vel_{(i,R1)} - 1, 0\right) \end{array} . \tag{7}$$

The previous local rule is a representation of a natural factor in the urban transit system and, in some way, can contribute to the rise in congestion. Another condition for the deceleration of the vehicle is the existence of obstacles on the road. The $nfree_i = gap(v_i)$ function is responsible for identifying the maximum number of free cells in which the vehicle can move in a given direction d, according to the Figure 1. The local rule for the downturn by obstacles is given by Equation 8.

$$(R3) : vel_{(i,R3)} = min\left(vel_{(i,R2)}, nfree_i\right) \tag{8}$$

The rule $R3$ simulates, to some extent, the vision of the driver, it means, the maximum that he can move is a combination of following factors: the road speed limit, maximum speed that the vehicle can reach and the next obstacle. Furthermore, it is defined in the model rules for local buses to consider the bus stops. Consider $S = \{(x_1, y_1), (x_2, y_2), \ldots, (x_n, y_n)\}$ the set of cells located in a bus stop. For a bus v_i, consider t_0 the moment where $loc_i \in S$ and k the stop duration, with a probability φ_i, defines de rules RS, as shown in Eq. 9.

$$\begin{array}{l} (RS): \\ If\ loc_i \in S,\ rand < \varphi_i,\ t < t_0 + k\ and\ v_i = bus \\ \quad vel_{(i,t+1)} = 0, \\ otherwise \\ \quad vel_{(i,t+1)} = vel_{(i,R3)} \end{array} . \tag{9}$$

Finally, the movement of the vehicle v_i given the direction d of displacement dir_i can be calculated by:

$$(R4) : loc_{i,t+1} = loc_i + vel_{i,t+1} * dir_i . \tag{10}$$

6 Overview on the Classification Methods

The traffic jam identification can be regarded as a problem of fault detection class, where the transit is the studied system. The identification can be binary (normal or congested) or multiple classes. In the following subsections will be a brief explanation of the methods used. For more details about used methods, we suggest to search the cited references.

6.1 Naïve Bayes Classifier

The NB is a simple and but efficient Bayesian network classifier. It is built upon the strong assumption that different attributes are independent with each other given the class [23]. Although this classifier has this strong assumption, studies show that its performance is not affected when the database does not have the attributes fully independent of each other [24]. Formally, the model for the classifier has the following form (using Bayes Theorem):

$$p(C|F_1,\ldots,F_n) = \frac{p(C)\ p(F_1,\ldots,F_n|C)}{p(F_1,\ldots,F_n)}. \tag{11}$$

Where $p(C)$ is the probability of occurrence of class C, $p(F_1,\ldots,F_n|C)$ is the maximum likelihood, and $p(F_1,\ldots,F_n)$ is the evidence. All these parameters can be obtained through the relative frequencies of training database [25].

6.2 Decision Tree Classifier

The DT is an inductive tree-like structure classifier where the basic idea is break up a complex decision into a union of several simpler decisions [26]. In the branch nodes of the tree some classification rules are stored. This is done in order to group similar samples in the same leaf nodes. DT is not, in general, the algorithm itself but a means to perform the classification. The best known algorithms to implement a DT are C4.5 and ID3. They differ, mainly, in the way of how the attributes are sequenced for the decision. The Figure 2 illustrates an example of DT with their *IF... THEN ...ELSE...* rules form.

6.3 K-Nearest Neighbor

The K-NN method is one of the most simplest and oldest classifier, but, at the same time, most important methods for regression and pattern classification. It is based on the fact that similar instances tend to be closer in search space. This method requires two parameters: k (which gives the method's name) and a metric d. The performance of this method of classification depends heavily

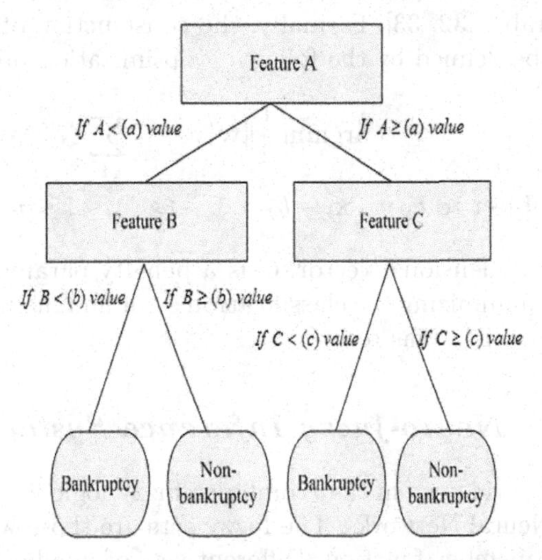

Fig. 2 An example of Decision Tree (source: Hung, C. and Chen, J.-H. (2009)).

on the metric applied. Whatever the metric used to it comply with the four properties below [27, 28]:

- No negativity: $d(a, b) \geq 0$;
- Reflexive: $d(a, a) = 0$;
- Symmetry: $d(a, b) = D(b, a)$;
- Triangular inequality: $d(a, b) + D(b, c) \geq d(c, a)$.

6.4 Artificial Neural Network

The ANN is a biologically inspired method capable of capturing highly complex non-linear functions. The fundamental unit of this network is called *neuron*, designed by McCulloch and Pitts. When many neurons work together to get a network called Multi-Layer Perceptron (MLP). There are several architectures of MLP as recurrent neural network (RNN), the self-organizings maps (SOM) and the radial basis function (RBF) where each one is capable of performing different tasks [29, 30]. The best known method of learning is called the backpropagation and is based on the motion made to correct the weights of each neuron, which is the exit to the entrance of the network [31].

6.5 Support Vector Machine

Initially created to linearly separable problems, the SVM was created by Vladimir Vapnik and co-authors in the late 90's. The basic principle behind SVM is to construct a hyperplane that is capable of separating the classes, where the distance (or the surface) between them is the maximum possible [29]. Recently, several methods were developed in order to adapt it for models not linearly separable [32, 33]. Formally, the construction of the hyperplane by the SVM can be defined by the following optimization problem [34]:

$$\text{argmin} \frac{1}{2}\|\mathbf{w}\|^2 + C\sum_i \xi_i \quad (12)$$

$$\text{subject to } c_i(\mathbf{w} \cdot \mathbf{x_i} - b) \geq 1 - \xi_i \quad 1 \leq i \leq n. \quad (13)$$

Where w is the n dimensional vector, C is a penalty parameter controls the trade-off between minimizing the classification error and maximizing the class separation margin, b is a bias term

6.6 Adaptive Neuro-fuzzy Inference Systems

The ANFIS is a hybrid system that combines fuzzy logic with learning abilities of Artificial Neural Networks. The fuzzy sets are those where each entry is associated to a member function. Different sets of numbers (where an element is or is not present), the fuzzy sets combine a degree of relevance to the

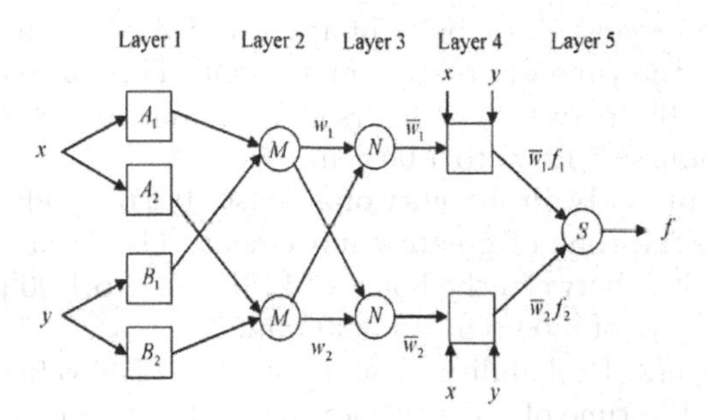

Fig. 3 An ANFIS architecture(source: Ubeyli, E. D. (2009))

element. The learning process of ANFIS is similar to an MLP, ie using back-propagation. The Figure 3 illustrates a typical architecture of ANFIS [35]. In brief, the Layer 1 associated with each entry a member's function. The layer 2 performs the multiply the degrees of the entries. The layer 3 is responsible for normalization of degrees. The layer 4 is responsible for defuzzyfication and finally, the layer 5 for the output.

7 Simulations and Results

7.1 Environment

The Figure 4 illustrates the layout of the implemented map to the simulator. It consists of 1 main (horizontal) and 3 via secondary (vertical) roads.

The main routes are composed of 3 lanes and its maximum allowed speed is 60 km/h (or 3 cells per iteration); furthermore, the secondary roads have only 2 lanes and maximum speed allowed is 40 km/h (or 2 cells per iteration). The entry of vehicles in the model is given in the following way:

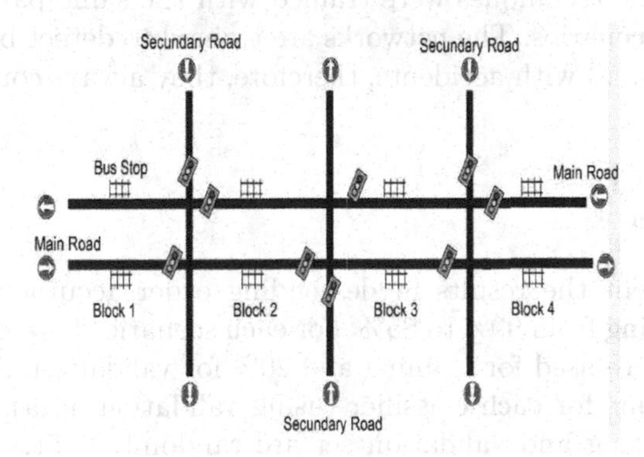

Fig. 4 Layout of implemented map of CAUTS

1. West-east Main roads: Probability of at least 10% of a vehicle entering the model outside the time of greatest movement. This probability increases linearly up to 70% between the hours of 7:00 a.m. to 8:00 a.m.. And, 50% between the hours of 12:00 to 1:00 p.m.
2. East-West Main roads: Probability of at least 10% of a vehicle entering the model outside the time of greatest movement. This probability increases linearly up to 50% between the hours of 12:00 a.m. to 1:00 p.m.. And, 70% between the hours of 4:00 p.m. to 5:00 p.m.
3. Secondary streets: Probability of at least 10% of a vehicle entering the model outside the time of greatest movement. Increasing 30% in the hours between 7:00 a.m. and 8:00 a.m., 12:00 and 1:00 p.m., and, 4:00 p.m. and 5:00 p.m..

For all scenarios are carried out 30% of large vehicles (between bus and trucks), and, all simulated accidents occurred on the central lane of the west-east main road, but in different blocks.

7.2 Parameters and Scenarios

The parameters used for the classifier are:

1. DT: was implemented using the C4.5 method, maximum depth=5;
2. K-NN: $k = 17$ and $d =$ *Euclidean distance*;
3. ANN: MLP neural network with four layers, being [2, 25, 25, 2] the number of neurons in each layer;
4. Fuzzy: Sugeno ANFIS using has 5 membership functions (Gaussian) for each entry;
5. SVM: $\xi = 0.5$;

We simulated 5 different scenarios (08:00 to 09:00) with situations: (i) without incidents; (ii) incident in the first block, (iii) incident in the second block, (iv) incident in the third block and, (v) incident in the fourth block, according to Fig. 4. The techniques were trained with the same parameters for all considered the scenarios. The networks are trained to detect between the situations without and with accidents, therefore, they always consider situation (i) as reference.

7.3 Results

Tables 1-4 present the results in descending order accuracy for all tested topologies, ranging from 99% to 85%. For each scenario there are 400 records, of which 80% were used for training and 20% for validation. Accuracy is the average of 35 runs for each classifier (using validation data). For each simulation the training and validation set are randomly split. The results for all tested topologies presented good accuracy. This is mainly due to the fact that a consistent (big enough) dataset can be arbitrary generated using the CAUTS model. Moreover, it appears that a breakdown in the first blocks is harder to detect than in the last ones. As it is well known, traffic jams propagates backwards, therefore, the information of the first sensor are richer than in the last ones. Indeed, more information is got when the accident takes place in last blocks. This empirical expectation is observed in Tables 1-4.

Table 1 Performance of classifiers considering the scenarios (i) x (ii).

Classifiers Performance	
Method	Accuracy
MLP	96.50%
SVM	95.17%
DT	92.46%
KNN	92.02%
NB	86.12%
ANFIS	85.68%

Table 2 Performance of classifiers considering the scenarios (i) x (iii).

Classifiers Performance	
Method	Accuracy
DT	96.66%
SVM	92.21 %
ANFIS	90.95%
MLP	88.78%
NB	88.13%
KNN	87.48%

Table 3 Performance of classifiers considering the scenarios (i) x (iv).

Classifiers Performance	
Method	Accuracy
MLP	99.87%
NB	94.38%
DT	94.37%
KNN	92.64%
ANFIS	90.39%
SVM	89.14 %

Table 4 Performance of classifiers considering scenarios (i) x(v).

Classifiers Performance	
Method	Accuracy
ANFIS	99.12%
MLP	97.34%
NB	93.68%
KNN	92.63%
SVM	91.67%
DT	89.15%

8 Final Considerations and Future Works

This paper has studied the use of machine learning techniques to detect car breakdowns in an urban environment. Measurements of traffic flow in several points in the main road are used to train the techniques. These measurements were simulated in our model called CAUTS. Using this simulator it is possible to generate several scenarios with low cost. Combining the tested methods in a voting machine will be explored in a future work. Additionally, this technique, which is based solely in the traffic flow, can be also combined with other ones, as ones based on computer vision. Indeed, detecting the traffic jams is one important aspect in the traffic flow control. Based on this detection, the traffic lights can be adjusted such a way to decrease the harsh caused by the breakdown. This is one of the future aspects to be explored in this work. In fact, it is important to improve both, the CAUTS model and the machine learning techniques.

Acknowledgment

The authors would like to thank CNPq, FAPEMIG and CAPES for the financial support.

References

1. Wolfram, S.: Statistical mecachics of cellular automata. In: Theory and Applications of Cellular Automata. World Scientic, Singapore (1986)
2. C. on Traffic Flow Theory and Characteristics. In: Traffic Flow Theory: A State-of-the-Art Report. Transportation Research Board / National Academy of Sciences (2001)
3. Papageorgiou, M.: Some remarks on macroscopic traffic flow modelling. Transportation Research A 32(5), 323–329 (1998)

4. Hoogendoorn, S.P., Bovy, P.H.: State-of-the-art of vehicular traffic flow modelling. Special Issue on Road Traffic Modelling and Control of the Journal of Systems and Control Engineering 215(4), 283–303 (2001)
5. Chen, S.H., Jakeman, A.J., Norton, J.P.: Artificial intelligence techniques: An introduction to their use for modelling environmental systems. Mathematics and Computers in Simulation, 379–400 (2008)
6. Mamei, M., Roli, A., Zambonelli, F.: Emergence and control of macro-spatial structures in pertubed cellular automata and implications for pervasive computing. IEEE Transactions on Systems, Man and Cybernatics - Part A, 337–348 (2005)
7. Rich, E., Knight, K.: Artificial Intelligence. McGraw-Hill, New York (1991)
8. Qi, Z., Boaming, H., Dewei, L.: Modeling and simulation of passenger alighting andboarding movement in beijing metro stations. Transportation Research Part C, 635–649 (2008)
9. Chen, C., Li, Q., Kaneko, S., Chen, J., Cui, X.: Location optimization algorithm for emergency signs in public facilities and its applications to a single-floor supermarket. Fire Safety Journal, 113–120 (2009)
10. Xie, D.-F., Gao, Z.-Y., Zhao, X.-M., Li, K.-P.: Characteristics of mixed traffic flow with non-motorized vehicles and motorized vehicles at an unsignalized intersection. Physica A: Statistical Mechanics and its Applications 388(10), 2041–2050 (2009)
11. Blue, V., Bonetto, F., Embrechts, M.: A cellular automata of vehicle self organization and nonlinear speed transitions. In: Proceedings of Transportation Reserach Board Annual Meeting, Washington, DC (1996)
12. Nagel, K., Schreckenberg, M.: Cellular automaton models for freeway traffic. Physics I (2), 2221–2229 (1992)
13. Schadschneider, A., Schreckenberg, M.: Cellular automaton models and traffic flow. Physics A (26), 679–683 (1993)
14. Villar, L., de Souza, A.: Cellular automata models for general traffic conditions on a line. Physica A (211), 84–92 (1994)
15. Nagel, K.: Particle hopping models and traffic flow theory. Physical Review E (3), 4655–4672 (1996)
16. Maerivoet, S., Moor, B.D.: Cellular automata models of road traffic. Physics Reports 419(1), 1–64 (2005)
17. Makowiec, D., Miklaszewski, W.: Nagel-schreckenberg model of traffic - study of diversity of car rules. In: International Conference on Computational Science, vol. (3), pp. 256–263 (2006)
18. Wahle, J., Neubert, L., Esser, J., Schreckenberg, M.: A cellular automaton traffic flow model for online simulation of traffic. Parallel Computing 27(5), 719–735 (2001)
19. Boccara, N., Fuks, H.: Critical behaviour of a cellular automaton highway traffic model. Journal of Physics A: Mathematical and General 33(17), 3407–3415 (2000)
20. Fukś, H.: Exact results for deterministic cellular automata traffic models. Phys. Rev. E 60(1), 197–202 (1999)
21. Nassab, K., Schreckenberg, M., Boulmakoul, A., Ouaskit, S.: Effect of the lane reduction in the cellular automata models applied to the two-lane traffic. Physica A: Statistical Mechanics and its Applications 369(2), 841–852 (2006)
22. Zhu, H., Lei, L., Dai, S.: Two-lane traffic simulations with a blockage induced by an accident car. Physica A: Statistical Mechanics and its Applications 388(14), 2903–2910 (2009)

23. Fan, L., Poh, K.-L., Zhou, P.: A sequential feature extraction approach for naïve bayes classification of microarray data. Expert Systems with Applications 36(6), 9919–9923 (2009)
24. Perez, A., Larranaga, P., Inza, I.: Bayesian classifiers based on kernel density estimation: Flexible classifiers. International Journal of Approximate Reasoning 50(2), 341–362 (2009)
25. Isa, D., Kallimani, V., Lee, L.H.: Using the self organizing map for clustering of text documents. Expert Systems with Applications 36(5), 9584–9591 (2009)
26. Safavian, S.R., Landgrebe, D.: A survey of decision tree classifier methodology. IEEE Transactions on Systems, Man, and Cybernetics 21(3), 660–674 (1991)
27. Choi, K., Singh, S., Kodali, A., Pattipati, K., Sheppard, J., Namburu, S., Chigusa, S., Prokhorov, D., Qiao, L.: Novel classifier fusion approaches for fault diagnosis in automotive systems. IEEE Transactions on Instrumentation and Measurement 58(3), 260–269 (2009)
28. Zuo, W., Zhang, D., Wang, K.: On kernel difference-weighted k-nearest neighbor classification. Pattern Anal. Applic. 11(3-4), 247–257 (2008)
29. Haykin, S.: Redes Neurais: Princípios e Prática. Bookman, vol. 2 (2004)
30. Delen, D., Fuller, C., McCann, C., Ray, D.: Analysis of healthcare coverage: A data mining approach. Expert Systems with Applications 36(2), 995–1003 (2009)
31. Braga, A.d.P., Carvalho, A.P.D.L.F.D., Ludemir, T.B.: Redes Neurais Artificiais - Teoria e Aplicações. LTC 2 (2007)
32. Schnell, S., Saur, D., Kreher, B.W., Hennig, J., Burkhardt, H., Kiselev, V.G.: Fully automated classification of hardi in vivo data using a support vector machine. NeuroImage 46(3), 642–651 (2009)
33. Maglogiannisa, I., Loukisb, E., Zafiropoulosb, E., Stasis, A.: Fully automated classification of hardi in vivo data using a support vector machine. Computer Methods and Programs in Biomedicine 95(1), 47–61 (2009)
34. Bae, M.H., Pan, R., Wu, T., Badea, A.: Automated segmentation of mouse brain images using extended mrf. NeuroImage 46(3), 717–725 (2009)
35. Jang, J.-S.R., Sun, C.-T., Mizutani, E.: Neuro-Fuzzy and Soft Computing. Prentice-Hall, Englewood Cliffs (1997)

Next Generation Environments for Context-Aware Learning Design

Patricia Charlton and George D. Magoulas

Abstract. Next generation Learning Design tools and applications have similar design requirements as intelligent applications that create, share and re-use content through the use of data specifications or formal models. In this paper, we present an approach that combines ontologies and autonomic computing principles to design and build next generation learning design environments that possess context-aware features. Our approach builds on the features of self-management and organisation of autonomic computing but uses self-configuration as a means to extend a knowledge-based inference through the design of meta-level inference. This leads to the design and implementation of a next generation learning design tool that is context-aware supporting both knowledge push and knowledge pull to enable appropriate use of theory and practice when creating learning designs for use in higher education.

1 Introduction

One of the current interests in the field of "Learning Design" is to find ways to support teachers who wish to develop designs that incorporate digital technologies [11]. The focus from pedagogical point of view is to enable teachers to exploit the constructivist potential of digital technologies for learning: those that support learners in discussing, collaborating, and creating user-generated designs.

The term "Learning Design" has been in use only in recent years; the earliest work in the field can be traced back to instructivist approaches, e.g. [10]. To make theoretical findings readily available to practitioners led to extensive work on Instructional Design Theory [15], which attempted to make learning theories more operational. The development of interest in "Learning Design" as a focus of research began with this recognition that the constructivist pedagogical theories did not easily transfer to the practice of teaching [13]. The emphasis on what learners were doing and how to support their activities was much less constrained

Patricia Charlton
London Knowledge Lab, Birkbeck College, University of London, UK
e-mail: patricia@dcs.bbk.ac.uk

George D. Magoulas
London Knowledge Lab, Birkbeck College, University of London, UK
e-mail: gmagoulas@dcs.bbk.ac.uk

I. Hatzilygeroudis and J. Prentzas (Eds.): Comb. of Intell. Methods and Appl., SIST 8, pp. 125–143.
springerlink.com

by constructivism. This dependence on the context in which learning takes place required an approach to teaching based on design principles rather than pre-defined instructional sequences [14]. Supporting these design principles has required re-thinking how to support learning designers.

By leveraging the semantic web developments and knowledge management, and exploiting the observation that knowledge management building blocks (ontological models) form the domain grounding for context-aware applications we have designed and implemented a framework for supporting next generation Learning Design (LD) tools. To manage and exploit the semantics of concepts used when creating the learning design we use self-configuration, an autonomic computing technique, which enables us to infer about appropriate context changes, as well managing context alignment via the underlying ontological models.

The paper is organized as follows: In Section 2 there is a review of learning design tools and identification of their limitations. Section 3 provides the requirements for a learning design environment and evaluates tools with respect to self-configurable and context-aware capabilities. Section 4 provides a short summary about the background of autonomic computing and context-aware systems and the use of ontologies. In Section 5 we present our approach to support context-aware learning design. Section 6 illustrates the overall architecture and self-configurable inference details demonstrating the creation and management of context-paths. Section 7 concludes the paper.

2 Learning Design Tools

Existing e-learning systems and authoring tools have several limitations in respect of support provided and usability, and cannot accommodate the needs of teachers who increasingly look for more intelligent services and support when designing instruction [12]. This support can be potentially helpful in formulating teaching goals and lesson plans and in better accommodating learners' needs by incorporating personalization technologies into teachers' designs. In fact, at present, systems do not provide tools for identifying patterns in effective practice and offer no opportunities for teachers to personalize the learning experience and collaborate with peers in developing more effective designs.

There is considerable work on developing various languages and formalisms for learning design (e.g. [22][23]). The Educational Modeling Language (EML), which appeared in 2000, was the outcome of work that started in 1997 by the IMS Global Consortium (IMS) and the OUNL. Initial work by the IMS targeted support processes for learning rather than the learning process itself but by early 2001 it was realized that a specification was needed to describe the learning processes. The EML approach to pedagogy is to provide a high-level abstraction of learning methods, including actors (e.g. tutors and students) and roles (e.g. activities) undertaken in an environment. The term "environment" has been used in this context to describe learning content, tools, communication, and other elements usable by learners and others in an activity. Activities are structured using a "learning flow" that includes decision-points (so that, for example, performance in one activity determines the next), sequences and choices. This high level of abstraction and flexibility makes EML a very powerful tool for expressing very different learning scenarios. The

EML focused on the entire learning process and was considered as complementary to the specifications developed by IMS. The IMS LD 1.0 adopted the XML format, which is not visible to the designer but works behind the scenes like the converters of document formats used in software applications (e.g. converting DOC format to HTML in MS Word).

Other attempts in this area internationally, include the PALO language (http://sensei.ieec.uned.es/palo/) and the E2ML [19]. The PALO approach allows creation of a course-specific repository of semantically linked material rather than a set of local or distributed knowledge objects, which leads to the construction of a knowledge base that is organized along a set of themes/learning scenarios. This is considered as a core aspect of course development in PALO. In E2ML, goals, requirements and design of the teaching and learning activities are described in a visual language. The E2ML model is compliant with the IMS LD specification; it can be integrated with Learning Object Metadata standards and its usability has been explored in several studies.

IMS LD has motivated developments in authoring using tools that exploit IMS LD concepts, such as the Unit of Learning (UoL), or are IMS LD compliant [20]. Some examples in this area are editors like CopperAuthor[1] and Cosmos[2] or the Reload LD Editor[3], which can be run together with other tools and engines, like CopperCore[4] or Sled[5]. However, current tools are not very friendly to non technical users as they assume that the teacher is familiar with the technical editors and the specifications. Paquette's work [22] uses OWL (the Web ontology language), as a key component in developing formal representations. This work can inform the development of next generation of environments for learning design by matching it with design-based representations that mesh with and extend effective teaching practice.

3 Rational and Overview of Our Approach

Although LD information can be quantified for engineering purposes, as done in the works mentioned above, only LD knowledge is of real social and economic importance and can help to increase the rate of adoption and change current teacher practices. This kind of knowledge must be assimilated by humans before they can use it and is not enough to copy information or reproduce learning design products mechanically using XML, Petri Nets or LAMS sequences. In our view a LD environment should work more like a system to manage tacit knowledge (i.e. knowledge acquired from practical experiences). This type of knowledge cannot be easily formalized (e.g. using Petri Nets) and not a single actor knows the whole picture. Thus it is hard to learn and pass on and this knowledge has not yet been given sufficient recognition in the approaches to LD so far. However, this form of knowledge is an essential part of the educational environment and affects its

[1] http://sourceforge.net/projects/copperauthor/
[2] http://www.collide.info/Members/admin/publications/ICCE05.260.pdf
[3] http://www.reload.ac.uk/
[4] http://coppercore.sourceforge.net/
[5] http://sled.open.ac.uk/

economic performance. Inclusion of the LD theory and practice from various perspectives and view points is the basis of our design which employs knowledge engineering to structure the LD concepts and learner designers' requirements for flexibility in the design process in order to provide an environment for enabling knowledge push. This knowledge is an important component of the teaching profession, and social cooperation and a common understanding are crucial to task performance.

Current LD tools described in Section 2 can be roughly organized into the following groups: (a) standards-based, (b) generic form-based, (c) authoring tools and (d) ontology-based. Table 1 provides an overview.

Table 1 Overview of Existing Tools and Main Properties

Properties	Learning Design Tools			
	Standards-based	*Generic form-based*	*Authoring*	*Ontology-based*
Self-configure	No	No	No	No
Context-aware	No	No	No	No
Inference	No inference about theory	No inference; static guidance about theory is used	Inference usually supports one theory	May provide inference
Formal ontology	No	In part	May support concepts about one theory	Yes, only a model or limited set of theory supported
Concepts	Concepts and schemas for usage and integration	Concepts about LD are available as part of the form	Yes	Yes

Standards-based approaches, such as Educational modeling languages and IMS-LD [2], provide greater interoperability between tools and designs. This approach enables building tools with specific functionality, e.g. LAMS[6], Moodle[7], which facilitate creating activity sequences, supporting from technical point of view modeling of various design methods, theories and approaches and generate designs through LD engines. Generic form-based tools, such as Phoebe[8] and CompendiumLD[9] are used for designing, managing and delivering learning activities and content, e.g. learning design documents and, in certain cases, enable collaboration, online learning and social networking (e.g. Cloudworks [10]). They

[6] http://www.lamsinternational.com/
[7] http://moodle.org/
[8] http://www.phoebe.ox.ac.uk/
[9] http://compendiumld.open.ac.uk/
[10] http://cloudworks.ac.uk/

focus on one aspect of the design process and as a result designers need to learn and engage with a number of tools to benefit from re-use. Authoring tools usually support a particular instructional design theory and employ inference engines that enable sequencing and presentation of instructional material depending on learner characteristics; an overview is given in [4]. Ontology-based tools represent domain concepts and relationships [5][6] as well as educational theories and relationships [3] among them. They facilitate communication and sharing of LD knowledge through common vocabularies and the development of a rational for learning designs use, modeling, and most of the time promote a particular pedagogy approach.

The tools discussed above do not support the context-aware needs for complex applications and they are limited in their adaptation to the context of use. While context-aware [9] is often classed as event-driven, here we mean data-driven or more precisely domain knowledge-driven: conventional context-aware systems use property values for matching and triggering related events while a knowledge-aware system makes inferences between related concepts based on a deep domain and knowledge understanding. In the Learning Design Support Environment (LDSE) presented in this chapter this is enabled by a domain ontology that defines relevant concepts, which form both part of the problem definition and a solution, and inference rules to determine concepts that relate to the user's current problem space. Thus, concepts, such as learning outcomes, learning activities, learning approaches etc. form the problem definition or the solution depending on the user's context of use, e.g. for a designer who is following an approach to LD that is organized in terms of particular learning outcomes, the LDSE would exploit relationships between sets of learning outcomes and types of activities that are defined in terms of learning approaches that best serve them.

In LDSE, thus, it is not data processing of properties and values in the conventional sense but concept relationships that influence the construction of a context path, which is then managed through LD inference rules to determine concepts that relate to the user's problem space, i.e. user's design requirements. Rules are triggered by LD concepts incorporated in users learning design and user actions during the process of LD. Moreover, this is an approach that does not make assumptions about a particular way of creating learning designs as typically done in LD tools described above; for example using standard templates or specific learning objects. Instead it is based on a context path that emerges as events, concepts and information become available.

Let us consider for example a designer who creates her learning design employing concepts represented in the LDSE ontology. As the designer selects concepts to create her learning design (such as the concept "Learning outcome") and expresses that the aim of her session is to "Communicate ideas in academically acceptable forms of expression and argument", the LDSE inference engine is determining those concepts that are appropriate for the particular context. To this end, LDSE builds upon a number of "Learning Approaches" that are available in the system supporting the various learning outcomes. For example, in this case, the learning outcome "Synthesis" (following Bloom's taxonomy) with instance "Communicate ideas" is supported by a learning approach "Collaborative learning" that is best served by a set of "Learning activities", whose instance "online collaborative project using simulator" could be suggested to the designer as a potentially useful element of her design. At any point the user can of course ignore specific concepts/suggestions of LDSE altering the context path.

Apart from ontologies and inference rules, our approach exploits the type of features of self-management and organization that is expressed in autonomic computing [1]. LDSE uses self-configuration as a means to extend its knowledge-based LD inference through the design of meta-level inference. The inclusion of a concept within a learning design means that other domain concepts maybe more relevant than they were before. A process of self-configuration at the meta-level permits the inference to inspect not just concepts but relationships as well. This is necessary in order to both manage and infer the creation of two context-paths. For example the addition to a learning design by the user can be a core concept (represented in the LDSE ontology), a modified concept, a shared concept that is situated, or content that is unknown to the LDSE system. The evaluation of the nature of the concept at that point in time by the inference engine prepares an alignment of concepts for next possible steps and LD tags. This provides a knowledge-aware application with flexibility in finding, using and presenting information to the user.

Lastly, our approach has been designed to support multiple theories about LD, which assists in creating application context. This has been developed based on interviews with LD practitioners and LD case studies (see [8] for a full description of the methods used to construct our approach) and is theoretically underpinned by the Conversational Framework (CF) [16]. CF provides conceptual depth and perspective round a number of pedagogical theories with a clear mapping of a unit of learning into the broader ideas of a constructivist perspective. This has led to an ontological design of the system, where a unit of learning may differ both in concept and content to work of others, such as Mizoguichi [12] who attempted to create a theory-aware environment by adopting a particular instructional design theory.

4 Autonomic Computing and Ontologies for Context-Aware LD

Autonomic computing is aimed at designing and building systems that are self-managing. The characteristics often attributed to an autonomic system are a self-managing, autonomous and ubiquitous computing environment that completely hides complexity, thus providing the user with an interface that exactly meets his/her needs [1]. While LD identified the need for flexibility to support the functionality of self-management, there are two common autonomic computing design challenges that should be addressed: (a) what context to use for self-management and (b) how to collect that particular context in both form and content.

Several researchers [7] [9] have tried to categorize context-aware applications according to subjective criteria; a taxonomy on context-aware features is proposed in [7]. There are three general categories of context-aware features that context-aware applications may support [9]: (a) services to a user, (b) automatic execution of a service, and (c) tagging of context to information for later retrieval.

Here we consider LDSE to be context-aware if it can extract, interpret and use context information and adapt its functionality to the current context of use. LDSE uses the ontologies to provide context to assist in the creation of learning designs on behalf of the user.

For example, let us consider the following scenario. The learning designer has chosen an annotated activity of the type "small discussion group". The system automatically constructs the context path for this learning design and offers to the user relevant "Learning Outcome" recommendations. The recommendations come constructed with content that can be edited or used directly. This inference is performed indirectly by the LDSE via the concept "Learning Approach". In LDSE, the LD concepts have many to many relationships, e.g. many activities may support the learning outcome "Comprehension". As the user constructs concepts to complete their learning design the LDSE is constructing relevant knowledge. When the user edits or shares this learning design the same knowledge can be used to reference appropriately the concepts and immediately provide different views of the knowledge, such as LDSE view, user view, modified view illustrating the changes, e.g. activity, class size support, learning experience view etc. The user can change properties and concepts, such as class size, use of activities, e.g. on-line resources versus face-to-face teaching to see immediately what the implication is for a particular design for both the learner and teacher. It is the system's perspective, built on the knowledge of the LD community. The learning designer is in control in terms of using recommendations when they seem appropriate and when to use system concepts rather than modified concepts etc. While terms can be changed, so can the context of use. Again, the system provides a "common context of use". The modifications are held as contextual preferences by the user. The same principles of knowledge-based and self-configuration are applied.

Fig. 1 illustrates the concepts used to automatically tag a learning design in the above scenario. This tagging is automatic as the inference engine is supporting the same concepts selected by the user that exist in the knowledge-base. Both the knowledge-base concepts and context of use are maintained in such a way that the inference is possible. This formal semantics underpin the design enabling preference re-use, e.g. such as previous terms or properties that have been created by the user and sharing of designs. Adaptation occurs using both the knowledge model and the self-configuration. The self-configuration principles build on the relationships and LD rules. This means that while self-configuration is generic, it is constrained by the principles of the domain. LD exploits the flexibility of concepts clusters and the modification of these concepts being constructed via the user-driven selection. The knowledge push and knowledge-aware is possible by exploiting the same principles of concept similarity, indirect mappings and look-ahead strategy of self-configurability.

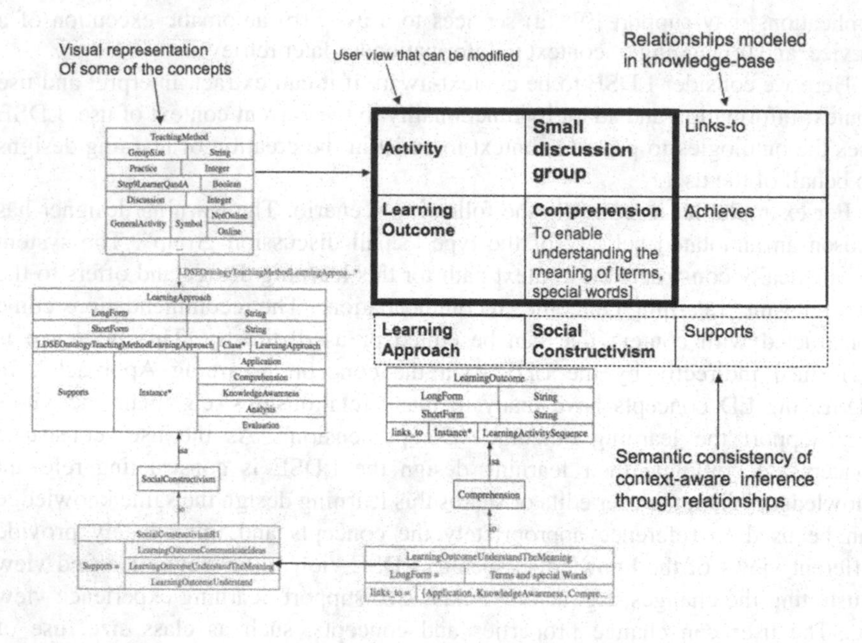

Fig. 1 Automatic creation of semantic tags for annotating learning design.

In LDSE, ontologies are used to enable the user to find learning designs that best match their LD approach: the user can enter LD concepts that exist in the knowledge-base. A user requiring assistance in adding learning outcomes to their design can request all known learning outcomes from the knowledge-base. The knowledge-base will use any context about the current learning design to appropriately present the learning outcomes. This is possible because the LDSE contains both concepts and relationships, thus defining a "concept context about LD". This is used as an inference template to assist in concept processing of the users request rather than a keyword search. Thus, the match is based on the formal processing of learning design concepts held in the knowledge-base.

The ontologies enable personalization methods to create and manage personal learning designs to be applied to the creation, search and retrieval of appropriate learning designs. They are also used to identify LD related concepts and content: the content entered by the user is related to the concepts in the knowledge-base, e.g. "Session topic" means that the user is working on concepts about a session. LDSE does not understand the domain of topics but the concepts about session. Hence as content is used within certain contexts, e.g. activities and session descriptions, the content is indirectly situated and thus tagged in this manner. This knowledge can be used to better categorize that content for further use.

The application of context is about how to maintain and enhance the context of tagging (see Fig. 1): the learning designs are triggering configuration rules that self-adapt and self-organize the underlying concepts of a learning design as changes take place by the user. The context, which is represented as a set of concepts that are used to build a context-path, can provide the relevant knowledge at run-time for the user. There are two context-based ontological models held by the LDSE. The first is the LDSE core ontology and the second is created at run-time when the user is creating their learning design, which may include modifications to the concepts that are part of the core ontology. As these modifications take place a meta-level inference is used to investigate the changes.

The self-configurable inference is about supporting "individual" contextual information, from history of use or declared alternative terms, or modified properties about learning design context. The individual context information can come from re-using previous learning designs, creation of personal terms from the common set of terms and refining the properties of common concepts that exist in the LDSE core ontology.

5 Self-configuration Approach and Use of Ontologies

In LDSE, users' individual terms and concepts support their personal preferences of creating learning designs. The context is gradually constructed as users enter LD concepts in their learning designs. LD concepts either refer directly to concepts in the core ontology or are, new concepts and content that are added by the user. Thus, they cannot be recognized as part of the core ontology and are "indirectly" referenced. The self-configurable context is built as a path. It is created as a conceptual network. Concepts and properties are matched in terms of similarity measure that operates as constraints in determining which cluster of concepts best reflect the creation of a learning design.

Self-configurable rules are used by the LDSE to assist in the automation of finding relevant concepts and content. The context of when to self-configure is built during the creation of the learning designs. If for example there is only one action to be taken in the next stage of the learning design, then self-configuration is simple – it is "execute" that action. If however, the user has made many modifications to the instances of a concept or set of concepts in use, e.g. editing a particular design. Then the LD engine needs to inspect the consequences of these changes both for appropriately tagging the design and for preparing possible next stages of the design. This requires the inference engine to reflect about if the current named concept is the most appropriately aligned concept. For example the user may have decided that the learning outcome they are designing for is *Application* (following Bloom's taxonomy). However, the user makes significant changes to the session and selects both activities and content material that from an LDSE perspective is for "Learning Outcome" *Evaluation*- to determine this inference requires the LDSE to inspect not just the properties of the concepts but the relationships that are either direct or indirect. This triggers a context-path alignment that is used to position further appropriate use of concepts. On the one hand the user's explicit reference to Application must be used as the user's label but on the other hand content retrieval of other design content is appropriately mapped to Evaluation. Application is still used but Evaluation content is also retrieved. Now if the user uses the Evaluation content then while the LD is tagged with Application externally, e.g. in the interface, the LDSE specific tagging does two things: first it tags with a user specific learning outcome Application but then has an LDSE alignment context path that tags this as

Evaluation. The result is when the learning design is shared or re-used it is used as an example of serving both Application and Evaluation learning outcomes. To achieve such alignment of concepts requires meta-level inspection that is a relationship inspection not just a concept instance or property inspection as to design the specific rules for every case would be extremely difficult. This also enables the user to create their own terminology that can still be underpinned by the LDSE, e.g. for example a user may create a learning outcome that is not part of Bloom's taxonomy. For example the user may wish to have a learning outcome called *Creativity*. Thus through the self-configuration the new learning outcome can be included and built. However, it does not form part of the core LDSE ontology but can be used and shared in the same way.

The similarity measure and rules of thumb define a set of conditions to be ideally satisfied. A particular priority order is provided by the LDSE but other configurations are enabled to permit user divergence while supporting the LD creation process - the "best support from the LDSE system" – as illustrated in the example given above about the user's view of learning outcome Application and LDSE view of the changes indicating learning outcome Evaluation. When there are user's preferences and concepts mapped to rules using concept alignment then the meta-inference is used to find a solution to: (a) create a context path, (b) link the appropriate core concepts, and (c) re-use of the formal semantic tagging (which is a representation of the context path created at run time).

5.1 Constructing a Context Path

When the LDSE receives a user input, it reasons about this input and sets up a context path that reflects its understanding of the learning design being created. This context is passed to the inference configuration and the necessary computations needed to deal with the input are created. Once a concept has been added to the context path then the management of the context-path requires handling further modification to properties and deletion of concepts. These changes require a call to "configure" an inference inspection by self-configure rules again with type similarity-context, or same, or indirect-context. However, it is possible that a configuration can result in a context-path being unbounded or highly fragmented. Essentially, the set of concepts changes made by the user are such that the mapping to the core ontology is limited, and thus limited knowledge can be inferred from the context-path related to the current learning design.

5.2 Managing the Context Path through Ontology Alignment

In the LDSE there are two context-paths created. The first path contains the original core LDSE ontologies and the second path the LDSE ontologies used, modified concepts and the user-created ontologies. Both context-paths form part of the formal semantic tagging of the learning design. This means that when the editing of the design happens, or sharing or re-use, the underlying concepts and the context in which they were used can be drawn from to support the creation of further learning designs. Fig. 1 provides an example of using the LDSE to create learning design using the ontology.

Within the LDSE the core ontologies are known, possible extensions are enabled and although general extensions are possible through the self-configurable approach the ontology extension is scoped. The creation and management of the context-path is considered a simplified form of ontology alignment, which is described as the task of finding relationships holding between the entities of two different ontologies; thus establishing a set of mappings expressing the correspondence between two entities of different ontologies through their relation.

The ontology alignment in LDSE is specialized to a set of correspondences between two ontologies, which are expressed as mappings between two context-paths as well as within a context path. The *mapping within a context-path* is a formal expression that states the semantic relation between two entities belonging to different ontologies. However, between the two context-paths this provides one ontology as an extension or modification of another and thus the similarity-measure becomes an evaluation of the extension or modification to use both the initial ontology and the user ontology edits to see if other core ontology may better serve to underpin the current learning design. The mappings are based on terminology and conceptual similarities. As the users are creating their learning design, some of the original concepts will be modified or new unknown concepts will be added. The concepts that are the same, modified or new are matched to the core ontology by using terminology and conceptual similarity. The result of this matching is used to generate the context-path of the learning design. Knowing the concepts that are similar is used to configure the next set of appropriate concepts for the user.

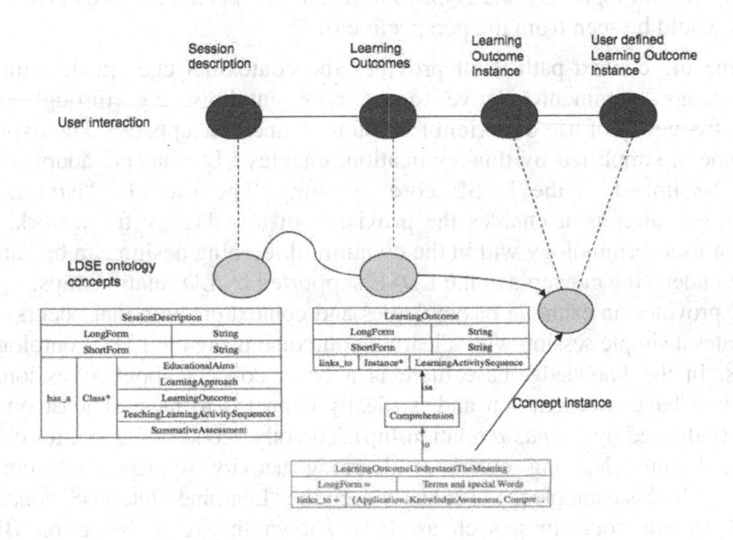

Fig. 2 An example of ontologies and context creation that occurs when a user creates a simple session.

Terminological similarity is the part of a mapping that expresses terminological relations between the lexical expressions used to name the entities to be mapped. Simple examples are: the name of two entities is the same, the name of an entity is an abbreviation of the name of the other or has been created by the user through the LDSE pedagogy thesaurus (see Section 6) and contains same properties but different names of the concepts.

Conceptual similarity is the part of a mapping that expresses the relation between entities in different ontologies. Simple examples are: concept c_1 in ontology O_1 is equivalent to concept c_2 in ontology O_2, concept c_1 in ontology O_1 is similar to concept c_2 in ontology O_2, instance i_1 in ontology O_1 is the same as instance i_2 in ontology O_2. O_2 is built as the user creates their learning designs and may never differ from the original ontology accept in values and terminology. However, extensions and modifications and deletion of properties are possible by the user and the inclusion of non core concepts means that handling the context of similarity and indirect mappings is required. This means:

- given two ontologies O_1 and O_2 *with different coverage*, tells us how the two ontologies can be used together to achieve a (less partial) description of the LD domain in LDSE. This permits clustering relevant concepts when the user is creating, editing or sharing a learning design.
- given two ontologies O_1 and O_2 *with different granularity*, tells us how facts in O_1 can be systematically translated into facts of O_2 (for example, how a fact f_1 belonging to O_1 can be rewritten as a logically equivalent fact f_2 in O_2). The user may chose a particular instance but wish to use a different set of terms. If the properties remain the same then the LDSE can draw from the original concepts and relationships to provide relevant information.
- given two ontologies O_1 and O_2 *with different perspective*, tells us how a fact f_1 in O_1 would be seen from the perspective of O_2.

Over time the context-path itself provides the contextual cues in determining a user ontology alignment relative to the core ontology, e.g. through indirect context, frequency of use of certain terminology and concepts etc. The expression of alignment, simplified by this application, enables LD concepts adapted by the user to be linked to the LDSE core ontology. The use of "*with different perspectives*" alignment enables the provision of a pedagogy framework, where the use of user terminology within the creation of learning design can be "aligned" with the underlying concepts in the LDSE supported by LD relationships.

Fig 2 provides an example of ontologies and context creation that occurs when a user creates a simple session with a learning outcome using the LDSE ontology core concepts. In the knowledge-base there is a basic concept about a session. Each session will have a description and is ideally expected to have at least one other concept (indicated by the *has-a* relationship). The other concept can be, for example, educational aims, learning outcomes, learning activity sequence or summative assessment. In this example the user has chosen the "Learning Outcome" concept and a set of learning outcomes, such as those shown in Fig. 3 based on Bloom's Taxonomy, are presented. Each learning outcome concept has a set of instances (an instance may also belong to a number of learning outcome concepts).

In Fig. 3 "a Learning Outcome" is an abstract concept, which has a relationship *links_to* "Learning Activity ". However, note the relationship from "Learning Activity" *achieves* "Learning Outcomes". For the expert the concept "Learning

Activity" has details of achieving a particular learning outcome. It is meant to be some type of content, tool, instruction that a teacher will use during the lesson. However, the relationship *links_to* from "Learning Outcome" to "Learning Activity" is intuitively more vague/loose relationship.

The concept hierarchy for "Learning Outcome" has a set of concrete classes that has the relationship "*isa*" (or *is a kind of*) e.g. "Comprehension" *is a kind of* learning outcome. Learning Outcomes also have relationships defined indirectly with session types and more directly with the learning approaches. Within certain concepts, properties are themselves concepts and so at any one interaction point with the user the LDSE inference engine can draw from the Knowledge-base (KB) a set of appropriate concepts to be used to help in the creation of learning design. This is the common use of KBs and inferencing to create inference-based contextual information. However, combining both the KB and self-configuration we can consider the broader scope of the LD context.

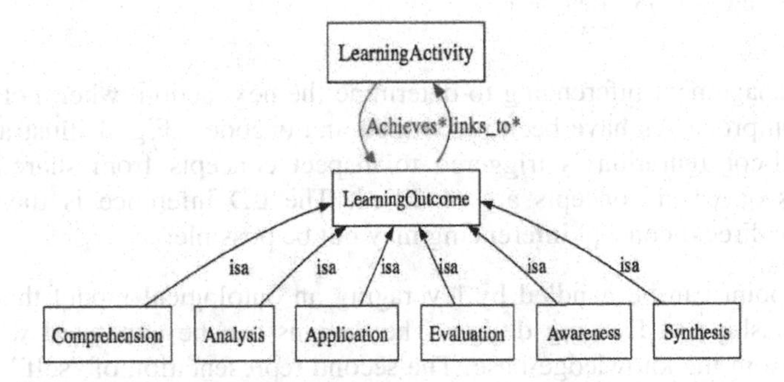

Fig. 3 Learning outcomes in the knowledge-base.

5.3 Self-configuration and Inference

Designs that include meta-interpreters and reflective techniques have been applied to enable the modeling of code as data that can be later included in the system execution. It is a highly compelling technique in distributed systems and is used in various ways to enable users to download and automatically install software, where configurations of the software to the hardware are possible. We use this technique in LDSE to enable self-configuration as part of the inference steps about the learning design application. The contextual knowledge gathered as a learning design is created can be used at each inference stage to select appropriate concepts. Within the LD context the concept of "self" can be coarsely divided into:

- The ability to handle high-level tasks and to automate the completion of these tasks. The possible types of knowledge are pre-defined (but not necessarily all instances of the knowledge) and the system has methods, rules and protocols to deal with the automation of tasks. In Fig. 4 the Learning Design Reasoner uses the LDSE ontology and user defined concepts to determine the concepts to be used and content inferencing for the user;

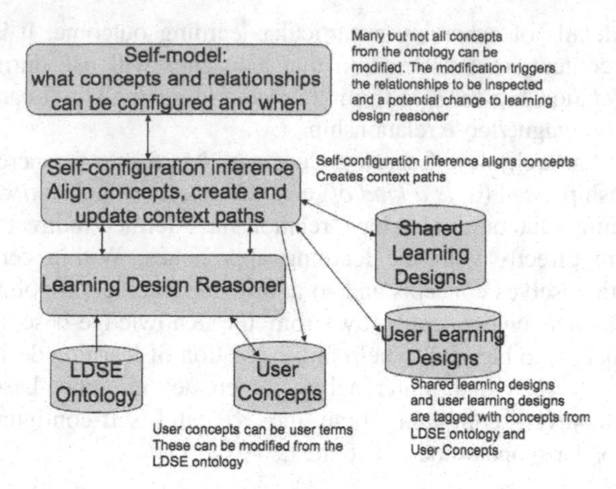

Fig. 4 Overview of LDSE Reasoner.

- Self-management inferencing to determine the next action, when not all of the decision processes have been predefined and encoded. Fig. 4 illustrates where the self-configuration is triggered to inspect concepts from shared learning designs or when concepts are modified. The LD inference is then effected because direc#t concept inferencing may not be possible.

The first point can be handled by leveraging an ontological model that captures the relationships of learning design. The designs can be annotated with formal concepts from the knowledge-base. The second representation of "self", where an inference has the possibility to incorporate self-management, is useful during the evaluation of a LD or when re-using LDs.

6 LDSE Architecture: Context-Awareness and Self-configuration Features

Fig. 5(a) and (b) provides an illustration of the main components of the LDSE. The conceptual model is represented by the LD ontology and includes relationships that support the use of pedagogy theory in practice. The pedagogy thesaurus permits the user to define their own terms that are linked to the original concepts found in the LD ontology. The usage of the new terms and change to properties triggers a contextual cue analysis to see if there are closer matching concepts than the original concept used. The user model and preferences relate to frequently used LD knowledge and re-use, such as concepts and terms from the pedagogy thesaurus or the LD ontology. The contextual information generated through the interaction with the system is stored keeping the context path details for later use, e.g. when editing a design. The learning designs that have been created are stored including the relationships and changes made from the original LD ontology, concepts developed in the pedagogy thesaurus and other contextual information from the user model and preferences. The learning designs are

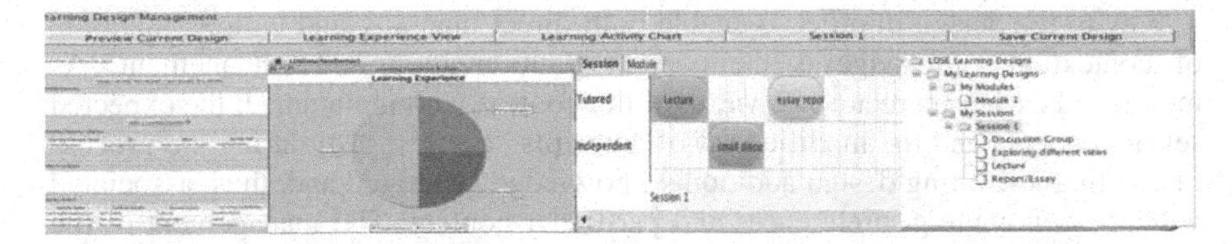

Fig. 5(a) Visualization and organization of the learning designs in LDSE 1.0.

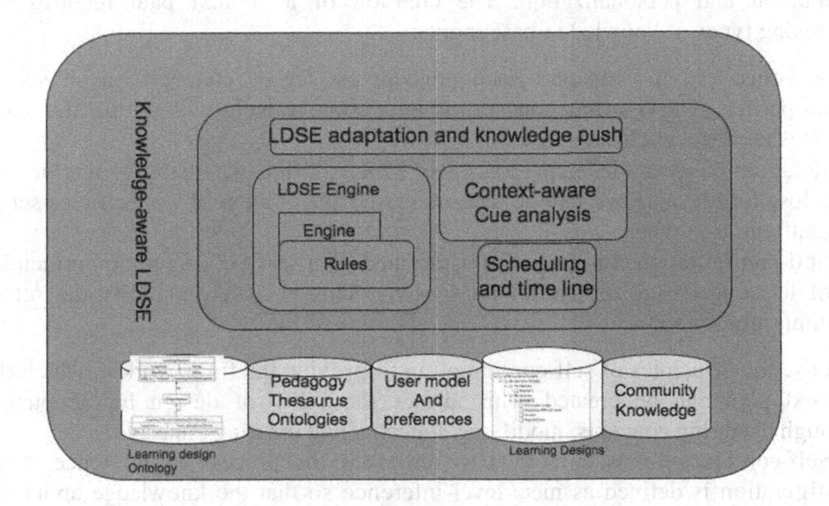

Fig. 5(b) Overview of context-aware LDSE architecture

automatically tagged by the system with the appropriated instances and modifications from the learning design ontology. The community knowledge contains the content to be shared and how this is shared and the re-use of other learning designs that contain the learning design tags.

The LDSE inference engine uses the ontology concepts to determine the appropriate knowledge, such as activities and learning outcomes to offer to the user. Contextual information is drawn from the user interactions and the conceptual models, learning designs and preferences held in knowledge-base. This permits "knowledge push" of recommendations to the user. The visualization of the designs is based on the same concepts from the ontology that have been modified to suite the learning design purpose. A time line presentation permits the user to see the set of scheduled activities. A pie chart permits the user to visualize the learning experience that is at the heart of their learning design, based on the concepts from the ontology. The form permits the user to enter free text and view recommendations and use the concepts that are available. The tree view of the learning design shows an overview to the user of the learning design, e.g. level of design, activities used during sessions etc.

Key to the design in order to enable both direct mapping and indirect mapping of contextual knowledge is the context path creation and management. As mentioned earlier there are two views of the context path maintained: the expected set of concepts and the modified set of concepts. There is also during the process of creating a learning design additional knowledge generated and, thus, associated with the context, e.g. preferences and particular content. The management of the context path is through the use of self-configuration system [24]. This enables both flexibility and context inspection at run-time; thus enabling the design adaptation and personalization. The creation of a context path requires the following types of knowledge-base inputs:

1. A collection of concepts. Each concept or set of concepts has a set of properties. The starting concept of any context path will define the most relevant concept clusters that will follow.
2. Relation between different concepts, which define the priority of concepts (close to/relevant) and can be determined by the context of a specific or set of configuration rules.
3. Self-configuration context, which is defined by a set of configuration principles of learning design as part of the system. This is underpinned by the set of configuration rules.

We use the principles of self-configuration to provide the framework in which the context path can be created. This allows flexibility of design for adaptation through changing concepts, modifying concepts and creating concepts.

Self-configuration is a method to represent the process of inference. Self-configuration is defined as meta-level inference so that the knowledge about the current context can be taken into account and appropriate concepts from the ontology can be included as certain data parameters change. The approach is used because it is impossible to design before hand all the choices and changes a user will make when creating their learning design. Very few properties of the LD concepts are fixed in terms of their values. While the LDSE has been designed with default preferences the true context and understanding of any given design is held in the user's mind. Thus, the inference process tries to align the changes made by the user with the relationships held in the knowledge-base.

In fact, what makes a concept unique is determined by the property, values and the relationships that concept has with other concepts. For example a teaching and learning activity has certain properties, e.g. an activity defined as "individual supervised project" means that the teacher's time with the student contains "individual supervision", i.e. the class size is one. If the activity is a group project then the class size is one or more. As the number of students change the learning experience through the group dynamics can change. Depending on the learning outcome there are different combinations and ways of organizing a session. It is possible during the design of a session that a teacher decides that the individual project supervision has a class size of five. There maybe many reasons that the teacher chooses to make this change. However, LDSE logically interprets this change as a trigger to find if other concepts and relationships internally are better suited to this new property value within the LDSE knowledge-base.

To this end, several meta-level partial inferences can occur. They are meta-level because to determine the alignment of the data the inspection must take place taking into account the set of possible alternative concepts. Once the LDSE has inspected and evaluated the possible alignments then normal first-order inference can continue. However, the normal first-order inference context is determined by the process of self-configuration about LD, to include new concepts and the relationship with other concepts. This process of suspending the process of reasoning, reasoning about the process, and using the results to control subsequent reasoning is called reflection [25]. For example, self-configure inference of LD inference about create LD context and a set of LDs, then create LD_i context produces a change to LD_i, if and only if current concept C_i of LD_i is different from the knowledge-base $LDSE_i$.

$$\text{Concept(create context(LD), LDSE)} \neq \text{Concept(LD, LDSE)}.$$

Once the change is recognized then self-configuration of a new alignment to be included within the creation of the LD occurs. The alignment uses the concepts (properties and values) and the relationships to determine what "influence" this change in the concept has created for this learning design.

The inferencing rules through self-configuration and the context path provide the necessary knowledge to manage two context paths and use this to generate semantic tags that take into account the user choices, modifications and original concepts from the domain. This permits both the creation and use of a pedagogy thesaurus, which is user-driven, and of user preferences. The preference model is built from frequency of use of the concepts and properties based on particular original concepts used and any modification of these concepts. Some of the concepts used by the user are "linked" by interaction only. That is the user generates a particular sequence to create their design. Other concepts are directly drawn from the domain concepts and are part of the initial LDSE context path. Other concepts are drawn from the pedagogy thesaurus, which links back to the original concepts. The preference concept may not link to the domain knowledge directly but is situated in a context of use and, thus, can be inferred through using the particular knowledge that is located near to the content. The nearness of a concept is determined by the context path and is based on the ontology alignment definitions given in Section 5. The knowledge is always driven by the original concepts but through the use of self-configuration the pedagogy thesaurus or preferences can be taken into account, and the creation of new or modified learning designs are adapted appropriately both with recommendations and automatic tagging by the system.

The current implementation of the LDSE, user interface and integration of the different components is in Java (see Fig. 5). Protégé has been used to develop the LD domain ontology and JESS has been used to implement the inference engine and core functions of the self-configurable framework and contextual cues.

7 Conclusion

Semantic web technologies and autonomic computing principles were combined in this paper in an attempt to design and build a next generation learning design environment. The paper described our approach which builds on the features of self-management and organization of autonomic computing using self-configuration as a means to extend a knowledge-based inference through the design of meta-level inference. Through leveraging the formal semantics of ontological models and inference techniques our approach illustrates some of the key cues for enabling context-aware computations that provide intelligent functionality. The system creates context paths, linking together domain concepts to keep track of the context in which user's learning designs are created. This leads to the design and implementation of a LD tool that is context-aware supporting both knowledge push and knowledge pull to enable appropriate use of theory and practice when generating learning designs for use in higher education. While being theory-aware is an important function of a learning design environment, being context-aware is also critical when including multiple resources and perspectives. In our approach these are combined through the inclusion of self-management functions as part of the knowledge inferencing. A preliminary evaluation of the context-aware features has been conducted with a small group of learning designers producing promising results for their effectiveness in supporting lecturers in practice.

References

[1] Steritt, R., Parasgar, M., Tianfield, H., Unland, R.: A Concise Introduction to Autonomic Computing. Advanced Engineering Informatics 19, 181–187 (2005)

[2] Koper, R.: Learning Design: A Handbook on Modeling and Delivering Networked Education and Training. Springer, Heidelberg (2005)

[3] Barn, B.S.: Conceptual Modelling of Educational Theories: An ontological approach. In: Proceeedings of the IADIS International Conference on Cognition and Exploratory Learning in Digital Age, pp. 45–51 (2006)

[4] Papanikolaou, K.A., Grigoriadou, M.: Building an Instructional Framework to Support Learner Control in Adaptive Education Systems. In: Magoulas, G.D., Chen, S.Y. (eds.) Advances in Web-based Education Personalized Learning Environments, IGI Publishing, pp. 127–146 (2006)

[5] Heiyanthuduwage, S.R., Karunaratna, D.D.: An Iterative and Incremental Approach for e-learning Ontology Engineering. International Journal of Emerging Technologies in Learning 4(1), 40–46 (2009)

[6] Armani, J., Botturi, L.: Bridging the Gap with MAID: A method for Adaptive Instructional Design. In: Magoulas, G.D., Chen, S.Y. (eds.) Advances in Web-based Education Personalized Learning Environments, IGI Publishing, pp. 147–177 (2006)

[7] Chen, G., Kotz, D.: A Survey of Context-Aware Mobile Computing Research. Technical Report TR2000-381, Department of Computer Science, Dartmouth College, Dartmouth Computer Science (2000)

[8] Charlton, P., Magoulas, G., Laurillard, D.: Designing for Learning with Theory and Practice in Mind. In: Proceedings of the Workshop "Enabling creative learning design: how HCI, User Modelling and Human Factors help", part of AIED 2009 Conference, pp. 52–61 (2009)

[9] Dey, A.K., Abowd, G.D.: Towards a Better Understanding of context and context-awareness. Georgia Institute of Technology, College of Computing, Technical Report GIT-GVU-99 22 (1999)

[10] Merrill, M.D., Twitchell, D. (eds.): Instructional Design Theory. Educational Technology Publication, Englewood Cliffs (1994)
[11] McNaught, C.: Identifying the complexity of factors in the sharing and reuse of resources. In: Littlejohn, A. (ed.) Reusing online resources – a sustainable approach to e-learning, pp. 199–211. London and Stirling, Kogan Page (2003)
[12] Mizoguchi, R., Bourdeau, J.: Using Ontological Engineering to Overcome AI-ED Problems. International Journal of Artificial Intelligence in Education 11(2), 107–121 (2000)
[13] Jonassen, D.H.: Thinking technology: Toward a constructivist design model. Educational Technology 34(2), 34–37 (1994)
[14] Oliver, R., Harper, B., Hedberg, J., Wills, S., Agostinho, S.: Formalising the description of learning designs. In: Goody, A., Herrington, J., Northcote, M. (eds.) Eds, Quality conversations: Research and Development in Higher Education, Jamison, ACT, HERDSA, vol. 25, pp. 496–504 (2002)
[15] Reigeluth, C.M.: What is Instructional-Design Theory and How Is It Changing? In: Reigeluth, C.M. (ed.) Instructional-Design Theories and Models: A New Paradigm of Instructional Theory, vol. 2, pp. 5–29. Lawrence Erlbaum Associates, Mahwah (1999)
[16] Laurillard, D.: Rethinking University Teaching. A conversational framework for the effective use of learning technologies. Routledge, London (2002)
[17] Mizoguchi, R., Hayashi, Y., Bourdeau, J.: Inside Theory-Aware Authoring System. In: Proceedings of the 5th International Workshop on Ontologies and Semantic Web for E-Learning (SWEL 2007), USA, July 9, 2007, pp. 1–18 (2007)
[18] Waterson, A., Preece, A.: Verifying ontological commitment in knowledge-based systems. Knowledge-Based Systems 12(1–2), 45–54 (1999)
[19] Botturi, L.: A Visual Language for Instructional Design: Evaluating the Perceived Potential of E2ML. In: Proceedings of EDMEDIA, Lugano, Switzerland (2010)
[20] Hernández-Leo, D., et al.: Reusing IMS-LD Formalized Best Practices in Collaborative Learning Structuring. Advanced Technology for Learning 2(4), 223–232 (2005)
[21] Naeve, A.: A SECI-based framework for professional learning processes. Deliverable 10.1 of the ProLearn EU/FP6 Network of Excellence 507310 (2007) IST 507310
[22] Paquette, G.: Graphical ontology modeling language for learning environments. Technology, Instruction, Cognition and Learning (TICL) 5, 2–3 (2007)
[23] Sicilia, M.-A.: Reuse, instructional design, and learning objects. In: Sicilia, M.-A. (ed.) Paper presented at Design-Based Approaches to Learning Objects and Learning Models Symposium Annual Conference of the American Educational Research Association (AERA), March 26, 2008, New York (2008)
[24] Charlton, P., Magoulas, G.: Self-configurable Framework for enabling Context-aware Learning Design. In: Proceedings of the IEEE International Conference on Intelligent Systems, pp. 1–6 (2010)
[25] Maes, P.: Computational Reflection., Ph.D Thesis, University of Brussels, Artificial Intelligence Laboratory (1987)

Neurules-A Type of Neuro-symbolic Rules: An Overview

Jim Prentzas and Ioannis Hatzilygeroudis

Abstract. Neurules are a kind of integrated rules integrating neurocomputing and production rules. Each neurule is represented as an adaline unit. Thus, the corresponding neurule base consists of a number of autonomous adaline units (neurules). Due to this fact, a modular and natural knowledge base is constructed, in contrast to existing connectionist knowledge bases. In this paper, we present an overview of our main work involving neurules. We focus on aspects concerning construction of neurules, efficient updates of neurule bases, neurule-based inference and combination of neurules with case-based reasoning. Neurules may be constructed from either symbolic rule bases or empirical data in the form of training examples. Due to the fact that the source knowledge of neurules may change with time, efficient updates of corresponding neurule bases to reflect such changes are performed. Furthermore, the neurule-based inference mechanism is interactive and more efficient than the inference mechanism used in connectionist expert systems. Finally, neurules can be naturally combined with case-based reasoning to provide a more effective representation scheme that exploits multiple knowledge sources and provides enhanced reasoning capabilities.

1 Introduction

The combination or integration of (two or more) different problem solving methods has given fruitful results in many application areas. The aim is to create combined formalisms or systems that benefit from each of their components. Disadvantages or limitations of specific intelligent methods can be surpassed or alleviated by their combination with other methods. It is generally believed that complex problems can be easier solved with such combinations (Medsker 1995).

Jim Prentzas
Democritus University of Thrace, School of Education Sciences, Dept. of Education Sciences in Pre-School Age, Nea Chili, 68100 Alexandroupolis, Greece
e-mail: dprentza@psed.duth.gr

Ioannis Hatzilygeroudis
University of Patras, School of Engineering, Dept. of Computer Engineering and Informatics, 26500 Patras, Greece
e-mail: ihatz@ceid.upatras.gr

I. Hatzilygeroudis and J. Prentzas (Eds.): Comb. of Intell. Methods and Appl., SIST 8, pp. 145–165.
springerlink.com

A popular type of combinations is that of symbolic and connectionist approaches, usually called the neuro-symbolic approach. Advanced neuro-symbolic formalisms and systems have been developed (Bookman and Sun 1993, Fu 1994, Medsker 1995, Hilario 1997, Sun and Alexandre 1997, McGarry et al. 1999; Wermter and Sun 2000, Cloete and Zurada 2000, d'Avila Garcez et al 2002, d'Avila Garcez et al 2004, Hatzilygeroudis and Prentzas 2004a, Bader and Hitzler 2005). Different types of neuro-symbolic approaches have been developed such as combinations of connectionist approaches with first-order logic (Bader et al. 2008, Shastri 2007), or with multi-valued logic (Komendantskaya et al. 2007) or with non-classical logic (d'Avila Garcez et al. 2007) or with symbolic rules (of propositional type) (Gallant 1993, Towell and Shavlik 1994, Fu 1993, Hatzilygeroudis and Prentzas 2000b and 2001b). However, combinations of neural networks and symbolic rules seem to have given more applied results (Souici-Meslati and Sellami 2006, Xianyu et al. 2008, Yu et al. 2008) due to the complementary advantages and disadvantages of the two combined formalisms (Hatzilygeroudis and Prentzas 2004a).

Symbolic rules have several advantages as well as some significant disadvantages in terms of knowledge representation and reasoning. Their main advantages involve naturalness of representation and modularity (see e.g. Reichgelt 1991). The naturalness of rules facilitates comprehension of their encompassed knowledge. Modularity refers to the fact that each rule is a discrete, autonomous unit enabling incremental development of the knowledge base as well as partial testing. Moreover, rule based systems provide an interactive inference mechanism, which guides the user in supplying input values, and an explanation mechanism, which justifies the reached conclusions. The provision of explanations is necessary in certain application domains (e.g. medicine) to justify system outputs. Symbolic rules have certain drawbacks besides advantages. An important disadvantage concerns the knowledge acquisition bottleneck that is, the difficulty in acquiring rules from experts (see e.g. Gonzalez and Dankel 1993). The brittleness of rules is another disadvantage. More specifically, it is not possible to draw conclusions from rules when there are missing values in the input data. For a specific rule, a certain number of condition values must be known in order to evaluate the logical function connecting its conditions. In addition, rules do not perform well in cases of unexpected input values or combinations of them.

Neural networks represent a totally different approach to problem solving, known as connectionism (see e.g. Gallant 1993, Haykin 2008). Neural networks possess certain advantages but disadvantages as well. They are able to obtain knowledge from training examples. Therefore, empirical knowledge (i.e. training examples) available in several domains is exploited and interaction with the experts is reduced. Additional advantages of neural networks concern their ability to generalize that is, provide computation of correct outputs from input combinations not present in the training set, their ability to represent complex and imprecise knowledge and their efficiency in producing outputs. Compared to symbolic rules, neural networks possess significant disadvantages. Main such disadvantages concern the lack of naturalness and modularity. It is difficult to comprehend the knowledge encompassed in neural networks and for this reason several rule extraction methods have been presented (Andrews et al. 1995). Due to the lack of modularity, a neural network cannot be decomposed into components and form a modular structure. The aforementioned drawbacks result into the difficulty (if not inability) in providing explanations for outputs produced by neural networks.

From the various neuro-symbolic approaches that have been presented, we concentrate on combinations that result in a uniform, seamless combination of the two component approaches. Such combinations are called unified, according to (Hilario, 1997), or integrated, according to (Bader and Hitzler, 2005). A main research direction at combining rules and neural networks involves use of prior domain knowledge in neural network configuration. One could discern two different trends in this research direction. The one trend stems from (Holldobler and Kalinke 1994), where a connectionist network is developed that implements the meaning function of a propositional (definite) logic program. The other trend stems from (Towell and Shavlik 1994), which consists of two main steps: an existing domain theory in the form of propositional rules is used to construct an initial neural network and then training data are used to train the network. On the other hand, connectionist expert systems are integrated systems that represent relationships between concepts associated with nodes in a neural network (Gallant 1988, Gallant 1993, Ghalwash 1998). The network also contains certain random cells that have no concepts assigned to them. These cells are introduced during construction.

Most (if not all) of existing such approaches give pre-eminence to connectionism. Thus, they do not exploit representational advantages of symbolic rules, like naturalness and modularity. Moreover, with the exception of connectionist expert systems, they do not provide the functionalities of a rule-based system, like interactive inference and explanation. It should also be mentioned that as far as connectionist expert systems are concerned, the presence of random cells results in certain incomprehensible explanations.

Neurules (Hatzilygeroudis and Prentzas 2000a, Hatzilygeroudis and Prentzas 2000b, Hatzilygeroudis and Prentzas 2001b) are a type of integrated rules combining symbolic rules (of propositional type) and neurocomputing. In contrast to other approaches, neurules give pre-eminence to the symbolic part of the integration. Therefore, they retain the naturalness and modularity of symbolic rules in a large degree. Neurules can be produced either from symbolic rules or from empirical data (Hatzilygeroudis and Prentzas 2000a, 2001b). Also a neurule-based system possesses an interactive inference mechanism (Hatzilygeroudis and Prentzas 2010) and provides explanations for drawn conclusions (Hatzilygeroudis and Prentzas 2001a). Mechanisms for efficiently updating a neurule base, given changes to its source knowledge (i.e. symbolic rules or empirical data), have also been developed (Prentzas and Hatzilygeroudis 2005, Prentzas and Hatzilygeroudis 2007b). Neurules may also be effectively combined with case-based reasoning (Prentzas and Hatzilygeroudis 2002, Hatzilygeroudis and Prentzas 2004c).

In this paper, we present an overview of our work concerning neurules. The structure of the paper is as follows. Section 2 presents the neurule-based knowledge representation scheme. In Section 3 production of neurules from existing symbolic rules is presented. Section 4 discusses aspects regarding the mechanism for efficiently updating a neurule base given changes to its symbolic source knowledge (i.e. symbolic rule base). Section 5 outlines construction of neurules from empirical data. Section 6 briefly discusses aspects regarding efficient updates of a neurule base due to availability of new empirical source data. Section 7 discusses the interactive neurule-based inference mechanism. Section 8 discusses issues concerning combination of neurules with case-based reasoning. Finally, Section 9 concludes.

2 Neurules

2.1 Syntax and Semantics

Neurules are a kind of integrated rules. The form of a neurule is depicted in Fig.1a. Each condition C_i is assigned a number sf_i, called its *significance factor*. Moreover, each rule itself is assigned a number sf_0, called its bias factor. Internally, each neurule is considered as an adaline unit (Fig.1b). The inputs C_i (i=1,...,n) of the unit are the conditions of the rule. The weights of the unit are the significance factors of the neurule and its bias is the bias factor of the neurule. Each input takes a value from the following set of discrete values: [1 (true), -1 (false), 0 (unknown)].

The output D, which represents the conclusion (decision) of the rule, is calculated via the standard formulas:

$$D = f(a), a = sf_0 + \sum_{i=1}^{n} sf_i C_i \qquad (1)$$

$$f(a) = \begin{cases} 1, & if \ a \geq 0 \\ -1, & if \ a < 0 \end{cases} \qquad (2)$$

where ***a*** is the *activation value* and *f(x)* the *activation function*, which is a threshold function. Hence, the output can take one of two values ('-1', '1') representing failure and success of the rule respectively. The significance factor of a condition represents the significance (weight) of the condition in drawing the conclusion. The LMS learning algorithm is used to compute the values of the significance factors as well as the bias factor of a neurule. Examples of neurules are shown in Table 3.

The general syntax of a neurule (in a BNF notation, where '< >' denotes non-terminal symbols) is:

<rule>::= (<bias-factor>) **if** <conditions> **then** <conclusion>
<conditions>::= <condition> | <condition>,<conditions>
<condition>::= <variable> <l-predicate> <value> (<significance-factor>)
<conclusion>::= <variable> <r-predicate> <value> .

where <variable> denotes a *variable*, that is a symbol representing a concept in the domain, e.g. 'sex', 'pain' etc in a medical domain, and <l-predicate> denotes a symbolic or a numeric predicate. The symbolic predicates are {is, isnot}, whereas the numeric predicates are {<, >, =}. <r-predicate> can only be a symbolic predicate. <value> denotes a value; it can be a symbol (e.g. "male", "night-pain") or a number (e.g "5"). <bias-factor> and <significance-factor> are (real) numbers.

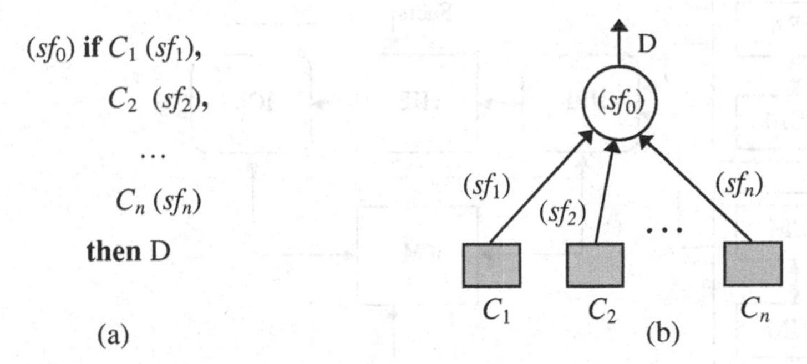

Fig. 1 (a) Form of a neurule (b) a neurule as an adaline unit

We distinguish three types of variables:

- *input or askable variables*, that is variables for which the user will be prompted to give a value during inference,
- *intermediate or inferable variables,* that is variables constituting intermediate goals of the inference process,
- *output or goal variables,* that is variables constituting the (final) goals of the inference process.

We also distinguish between *input, intermediate* and *output neurules.* An input neurule is a neurule having only input variables in its conditions and intermediate or output variables in its conclusions. An intermediate neurule is a neurule having at least one intermediate variable in its conditions and intermediate variables in its conclusions. An output neurule is one having an output variable in its conclusions.

2.2 *Neurule-Based System Architecture*

In Figure 2, the architecture of a neurule-based system is illustrated. The run-time system (in the dashed rectangle) consists of five modules: the *neurule base (NRB)*, the *hybrid inference engine (HIE)*, the *working memory (WM)*, the *explanation mechanism (EXM)* and the *indexed case library (ICL)*. The first four of these modules are more or less functionally similar to those of a conventional rule-based system. HIE in combination with ICL can provide additional reasoning capabilities (i.e. handling of exceptional situations).

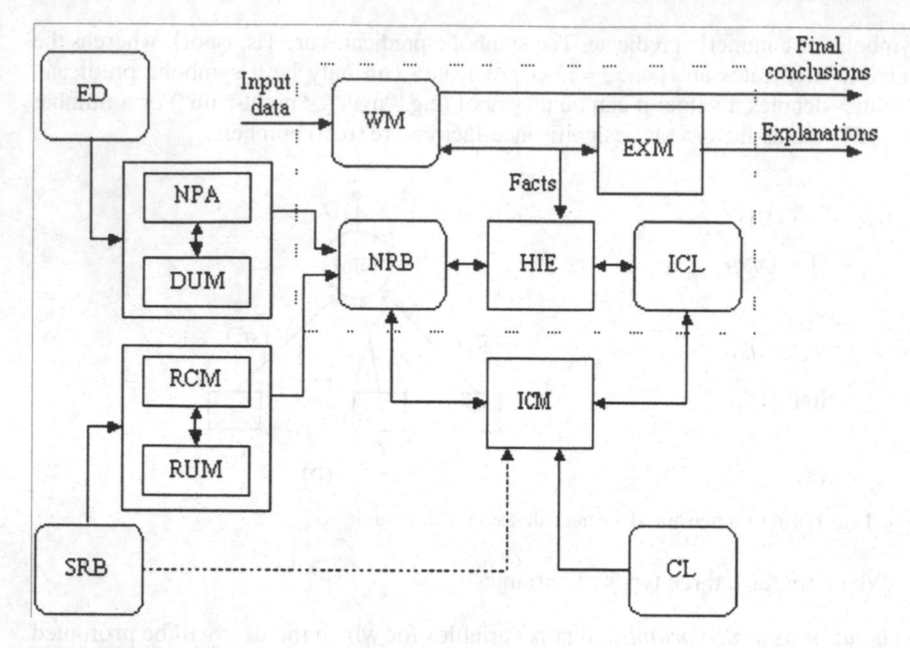

Fig. 2 The architecture of a neurule-based system

NRB contains neurules alongside certain information useful for updating neurules when the source knowledge changes (see Sections 4 and 6). HIE is responsible for making inferences. HIE either performs purely neurule-based inference by taking into account the data in WM and the neurules in NRB or combines neurule-based with case-based reasoning by also taking into account cases stored in the ICL. WM contains *fact assertions* either given by the user, as initial input data or during an inference course, or produced by the system, as intermediate or final conclusions during an inference course. ICL contains cases indexed by neurules in the NRB and is used by the approach combining neurule-based with case-based reasoning (see Section 8).

The architecture also includes certain offline modules useful for producing and updating the contents of the NRB and for constructing an *indexed case library (ICL)*. The contents of the NRB are produced from a *symbolic rule base (SRB)* or from *empirical data (ED)*. Construction of a NRB from a symbolic rule base is performed by the *rule conversion mechanism (RCM)* presented in Section 3. Construction of a NRB from empirical data is performed by the neurules production algorithm (NPA) presented in Section 5. The *rule update mechanism (RUM)* updates the NRB to reflect changes to its symbolic rule source (see Section 4). RUM interacts with the RCM to perform its tasks. The *data update mechanism (DUM)* updates the NRB when new empirical data becomes available (see Section 6). The *indexing construction mechanism (ICM)* constructs an ICL by taking as input a *case library (CL)* and either symbolic rules indexing cases in CL or neurules.

3 Construction of a Neurule Base from a Symbolic Rule Base

As mentioned above, neurules can be produced from either symbolic rules or empirical data. Here, we concentrate on the former.

An existing (propositional type) SRB can be converted to a neurule base (NRB) by the rule conversion mechanism. The corresponding conversion mechanism is described in (Hatzilygeroudis and Prentzas 2000b). Conversion does not involve refinement of SRB, but creates an equivalent knowledge base. This means that the conclusions drawn from NRB are the same as those drawn from SRB, given the same inputs. Each produced neurule usually merges two or more symbolic rules with the same conclusion. Therefore, the size of the produced NRB is less than that of SRB as far as both the number of rules and the number of conditions is concerned. This results in improvements to the efficiency of the inferences from NRB, compared to those from SRB (Hatzilygeroudis and Prentzas 2000b).

The conversion mechanism is outlined as follows:

1. Group symbolic rules into merger sets.
2. From each merger set, produce a merger.
3. Produce a training set for each merger.
4. Train each merger and produce one or more neurules.

Each *merger set* contains all the rules of the SRB having the same conclusion. We call such merger sets *initial merger sets*. A *merger* is a neurule having as conditions all the conditions of the symbolic rules in the corresponding merger set (without duplications) and significance factors as well as bias factor set to zero (or any other proper initial value). Each training set is extracted from the truth table of the combined logical function of the rules in its merger set (the disjunction of the conjunctions of the conditions of each rule), via a filtering process. Filtering eliminates the invalid rows of the truth table. Invalid rows are those with contradicting or inconsistent values.

Training of mergers is performed using the standard LMS algorithm. A limitation of the LMS algorithm is its inability to find a set of significance and bias factors that classify correctly all of the training patterns, in case that the training patterns of the training set are inseparable. In case that training is successful, one neurule will be produced. Otherwise, a splitting process is followed, which produces more than one neurule having the same conclusion, called sibling neurules.

Splitting is performed in a way that each subset contains symbolic rules that are 'close' to each other in some degree. *Closeness* between two symbolic rules is defined as the number of their common conditions. Splitting is based on the notion of closeness due to the observation that separable sets have rules with larger average closeness than inseparable ones. A *least closeness pair (LCP)* of rules in a merger set is a pair of rules with the least closeness (LC) in the set. Initially, a LCP in the merger set is found and two subsets are created each containing as its initial element one of the rules of that pair, called its pivot. Each of the remaining rules is distributed between the two subsets based on their closeness to their pivots. That is, each subset contains rules, which are closer to its pivot. If training fails, for a merger of a merger subset, the corresponding subset is further split into two other subsets, based on one of its LCPs. This continues, until training succeeds or the merger subset contains only one rule that is converted into a neurule.

Table 1 An example merger set

R1 **if** patient-class is human0-20, pain-feature2 is continuous, fever is no-fever, antinflam-reaction is none **then** disease-type is primary-malignant	**R5** **if** patient-class is human0-20, pain-feature2 is night, fever is no-fever, antinflam-reaction is none **then** disease-type is primary-malignant
R2 **if** patient-class is human0-20, pain-feature2 is night, fever is low **then** disease-type is primary-malignant	**R6** **if** patient-class is human21-35, pain-feature2 is night, antinflam-reaction is none **then** disease-type is primary-malignant
R3 **if** patient-class is human0-20, pain-feature2 is night, fever is medium **then** disease-type is primary-malignant	**R7** **if** patient-class is human36-55, pain-feature2 is night, fever is low **then** disease-type is primary-malignant
R4 **if** patient-class is human0-20, pain-feature2 is night, fever is high **then** disease-type is primary-malignant	**R8** **if** patient-class is human36-55, pain-feature2 is night, fever is medium **then** disease-type is primary-malignant

As an example, to demonstrate application of the main steps of the conversion mechanism, we use the merger set of Table 1 that consists of eight symbolic rules {R1, R2, R3, R4, R5, R6, R7, R8}, taken from a medical diagnosis rule base. The merger constructed from this (initial) merger set contains the ten distinct conditions of the eight rules and is shown in Table 2. The training set of the merger is extracted from the truth table of the combined logical function of the rules of the merger set:

$$\begin{aligned} F = {} & (C1 \wedge C2 \wedge C3 \wedge C4) \vee (C1 \wedge C5 \wedge C6) \vee (C1 \wedge C5 \wedge C7) \\ & \vee (C1 \wedge C5 \wedge C8) \vee (C1 \wedge C5 \wedge C3 \wedge C4) \vee (C9 \wedge C5 \wedge C4) \\ & \vee (C10 \wedge C5 \wedge C6) \vee (C10 \wedge C5 \wedge C7) \end{aligned}$$

where C1≡patient-class is human0-20, C2≡pain-feature2 is continuous, C3≡fever is no-fever, C4≡antinflam-reaction is none, C5≡pain-feature2 is night, C6≡ fever is low, C7≡ fever is medium, C8≡ fever is high, C9≡ patient-class is human21-35, C10≡ patient-class is human36-55.

Table 2 The merger of the merger set of Table 1

(0) **if** patient-class is human0-20 (0),
pain-feature2 is continuous (0),
fever is no-fever (0),
antinflam-reaction is none (0),
pain-feature2 is night (0),
fever is low (0),
fever is medium (0),
fever is high (0),
patient-class is human21-35 (0),
patient-class is human36-55 (0)
then disease-type is primary-malignant

The truth table of F contains 2^{10}=1024 training patterns, from which only 120 patterns remain after application of the filtering process. The training patterns of the training set are inseparable and the initial merger set is split in two subsets: MS1={R1, R5, R6} and MS2={R2, R3, R4, R7, R8}. The LCP that guides splitting is (R1, R7). Training of the merger of MS1 is not successful. So, {R1, R5, R6} is split in {R1, R5} and {R6} with LCP: (R1, R6). The merger of {R1, R5} is successfully trained and neurule NR1-5 is produced. Rule R6 is converted to a neurule (i.e. NR6). The merger of MS2 is successfully trained and neurule NR2-3-4-7-8 is produced. So, finally, from the initial merger set of eight symbolic rules, three neurules are produced. The produced neurules are shown in Table 3.

Table 3 Neurules produced from the merger set of Table 1

NR1-5
(-2.5) **if** fever is no-fever (1.4),
antinflam-reaction is none (1.3),
patient-class is human0-20 (0.8),
pain-feature2 is continuous (0.8),
pain-feature2 is night (0.8)
then disease-type is primary-malignant

NR2-3-4-7-8
(1.6) **if** patient-class is human0-20 (8.5),
pain-feature2 is night (8.2),
fever is medium (8.2),
patient-class is human36-55 (5.0),
fever is low (4.4),
fever is high (0.8)
then disease-type is primary-malignant

NR6
(-2.4) **if** patient-class is human21-35 (1.5),
pain-feature2 is night (1.4),
antinflam-reaction is none (1.3)
then disease-type is primary-malignant

4 Efficient Updating of a Neurule Base Produced from a Symbolic Rule Base

An aspect of interest involves the efficient updates of a NRB to reflect changes to its symbolic source knowledge. The basic changes to SRB can be (a) insertion of a new rule and (b) removal (or deletion) of an existing rule, since modification of a rule is equivalent to removal of the old rule and insertion of the new one. One approach to reflect such changes would be to reconvert the whole SRB and reproduce the whole NRB. Obviously such an approach would impose useless computational effort due to the fact that only specific parts of the SRB are affected from changes. To minimize the computational effort for performing updates, an efficient mechanism has been developed (Prentzas and Hatzilygeroudis 2005) that reconverts as small portion of SRB as possible. The modularity of NRB enables such an approach. Furthermore, the number of neurules after an update remains as small as possible, which is a significant aspect in terms of inference time-efficiency.

The update mechanism exploits the structure of a tree, called the splitting tree, that stores information related to the conversion process. More specifically, a splitting tree is used to store the splitting process for each initial merger set. The root of a tree corresponds to an initial merger set. The intermediate nodes and leaves correspond to the subsequent subsets, into which the initial merger set was split. An intermediate node denotes a subset that was split, due to training failure, whereas a leaf denotes a subset that was successfully trained and produced a neurule. The pivot of each (sub)set is attached to the corresponding branch of the tree. Figure 3 depicts the splitting tree corresponding to the splitting process for the merger set of Table 1.

Whenever a new symbolic rule R is inserted in SRB and there are more than one sibling neurule of R in NRB, the splitting tree is exploited to focus the update process on the neurules produced from the merger subset containing the rules closest to R. To achieve this, the splitting tree is traversed, starting from the root. Traversing is based on the closeness of the inserted rule to the LCP members of the merger subsets corresponding to the traversed nodes. Traversing ends at an intermediate node, when the corresponding merger subset contains a rule R' whose closeness to the inserted rule is less than the least closeness. Otherwise, traversing ends at a leaf. In case traversing stops at a leaf, the corresponding neurule is removed from the NRB, the merger corresponding to the new merger set is trained, updating accordingly the NRB and the splitting tree. In case traversing stops at an intermediate node, the descending nodes as well as corresponding neurules are removed, the new merger set is split in two subsets based on LCP (R, R'), the two corresponding mergers are trained updating accordingly the NRB and the splitting tree. In any case, parts of the initial splitting tree are exploited to avoid useless computational effort.

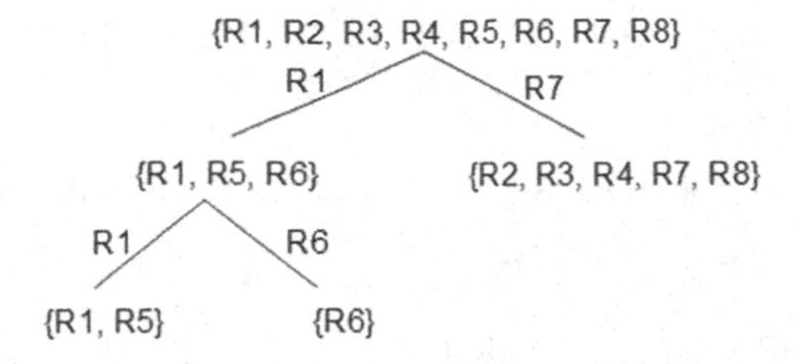

Fig. 3 The splitting tree for merger set of Table 1

Table 4 Inserted rule R9 and resulted neurule NR2-3-4-7-8-9

R9	**NR2-3-4-7-8-9**
if patient-class is human36-55, pain-feature2 is night, fever is high **then** disease-type is primary-malignant	(1.6) **if** pain-feature2 is night (8.2), fever is high (8.0), patient-class is human36-55 (5.0), patient-class is human0-20 (4.9), fever is medium (4.6), fever is low (4.4) **then** disease-type is primary-malignant

Whenever an existing symbolic rule R is removed from SRB and there are more than one sibling neurule of R in NRB, the splitting tree is exploited in a way similar to the approach for rule insertion. Traversing ends at an intermediate node, when R is a member of LCP of its merger (sub)set. Otherwise, traversing ends at a leaf. Each case is handled accordingly.

It should be mentioned that in certain situations of rule insertion/removal, the number of neurules contained in the NRB may decrease by one. A detailed presentation of the update mechanism along with experimental results is presented in (Prentzas and Hatzilygeroudis 2005).

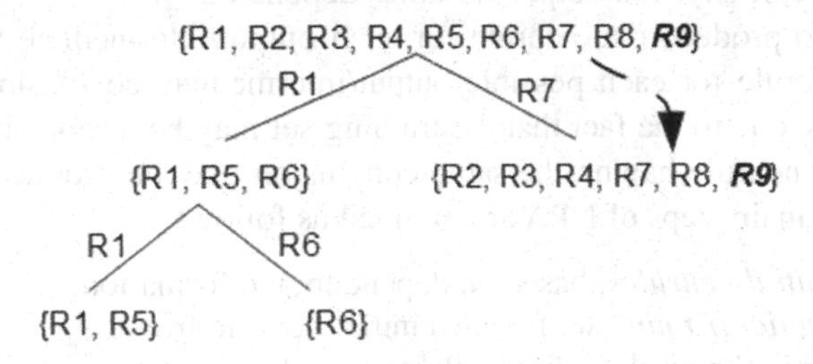

Fig. 4 Insertion of R9: traversal of the splitting tree

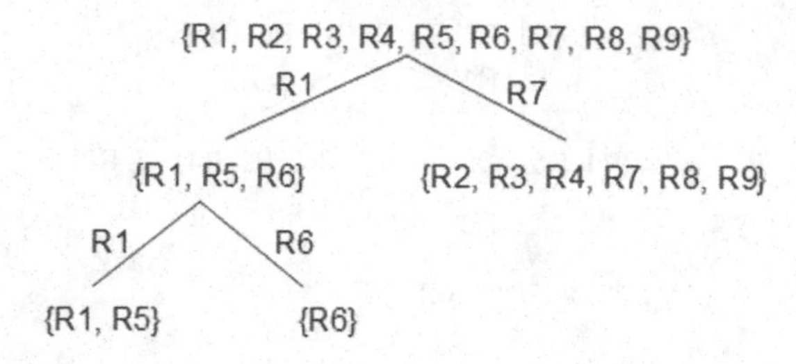

Fig. 5 Final form of the splitting tree after insertion of R9

We will demonstrate application of the update mechanism for rule insertion with an example. Let us consider the rules in Table 1 as constituting SRB and those in Table 3 as constituting NRB. Also, suppose that rule R9 (Table 4) is to be inserted. Information contained in the splitting tree shown in Figure 3 is exploited to efficiently perform the update of the NRB. Traversing of the splitting tree ends at the leaf related to subset {R2, R3, R4, R7, R8} (Figure 4). Notice that R9 is inserted into the merger sets corresponding to all traversed nodes. NR2-3-4-7-8 is removed from NRB. Training of the merger corresponding to the new merger (sub)set {R2, R3, R4, R7, R8, R9} is successful and the corresponding neurule NR2-3-4-7-8-9 is inserted into NRB (Table 4). The splitting tree takes the form shown in Figure 5.

5 Producing a Neurule Base from Empirical Data

In several domains, empirical data in the form of training examples are available and can be exploited to construct neurule bases. The neurules production algorithm (NPA) constructs a neurule base from empirical data. NPA requires the following input: a set of domain variables V representing the domain concepts with their possible values, possible dependency information among domain variables and a set of empirical data S. Dependency information indicates which variables the intermediate, if any, and output variables depend on.

NPA tries to produce one neurule for each output/intermediate variable value that is, one neurule for each possible output/intermediate conclusion. This is not always possible due to the fact that the training set may be inseparable. Therefore, more than one neurule having the same conclusion may be produced (i.e. *sibling neurules*). The main steps of NPA are outlined as follows:

1. Construct *initial neurules*, based on dependency information.
2. Extract an *initial training set* for each initial neurule from S.
3. Train each initial neurule individually and produce corresponding neurule(s).

Initial neurules represent the possible intermediate or final conclusions. One initial neurule is constructed for each value of each intermediate or output variable. The conditions of each initial neurule include the variables that contribute in drawing the corresponding conclusion, as specified by the dependency information. Then, for each initial neurule its corresponding initial training set is extracted from the empirical dataset. A training pattern has the form [v_1 v_2 . . . v_n d], where d is the desired value of a variable related to an intermediate or output conclusion and v_i, i=1,…,n are the values of the variables it depends on, called component values. We distinguish between *success examples* and *failure examples* in a training set. Success examples are those having '1' ('true') as their desired value, whereas failure examples are those having '-1' ('false'). Each initial neurule is individually trained, via the Least Mean Square (LMS) algorithm, using its own training set. Training is not always successful, that is a set of significance and bias factors cannot always be found that correctly classify all of the training examples. This is the case when the training patterns are inseparable. When the algorithm succeeds, that is values for the bias and significance factors are calculated that classify all training patterns, one neurule is produced. When it fails, due to inseparability of the training examples, a splitting process is followed. More specifically, the initial training set of the neurule is split into two subsets and two copies of the initial neurule are trained, each using one of the training subsets. If training of either neurule copy fails, its subset is further split into two other subsets and so on, until there is no failure or a subset contains only one success pattern. In this way, more than one neurule are produced, having the same conditions with different bias and significance factors and the same conclusion.

Splitting is based on the notion of closeness between training patterns. The closeness between two examples is defined as the number of their common component values. A *least closeness pair (LCP)* consists of two success examples that have the least closeness between them. Splitting a training set is based on an LCP. More specifically, each subset comprises one of the members of an LCP, the success examples closer to it and all the failure examples of the initial training set. This stems from the intuition that existence of quite different examples causes inseparability.

To demonstrate application of NPA, we use an example problem taken from the UCI Machine Learning ftp repository (Frank and Asuncion 2010); it is called the LENSES problem. There are five domain variables, four input (i.e. age, spectacle, astigmatic, tear-rate) and one output (lenses-class) that depends on the four input variables. Table 5 shows the corresponding empirical dataset consisting of twenty-four (24) patterns. The output variable takes three possible values (i.e. no-lenses, soft-lenses, hard-lenses) and therefore three initial neurules are constructed. A training set is extracted for each initial neurule. Each of the three training sets consists of twenty-four (24) patterns. The patterns in each of them have the same input values, but different output values.

Table 5 The empirical data set S for the lenses example problem

age	spectacle	astigmatic	tear-rate	lenses-class
young	myope	no	reduced	no-lenses
young	myope	no	normal	soft-lenses
young	myope	yes	reduced	no-lenses
young	myope	yes	normal	hard-lenses
young	hypermetrope	no	reduced	no-lenses
young	hypermetrope	no	normal	soft-lenses
young	hypermetrope	yes	reduced	no-lenses
young	hypermetrope	yes	normal	hard-lenses
pre-presbyopic	myope	no	reduced	no-lenses
pre-presbyopic	myope	no	normal	soft-lenses
pre-presbyopic	myope	yes	reduced	no-lenses
pre-presbyopic	myope	yes	normal	hard-lenses
pre-presbyopic	hypermetrope	no	reduced	no-lenses
pre-presbyopic	hypermetrope	no	normal	soft-lenses
pre-presbyopic	hypermetrope	yes	reduced	no-lenses
pre-presbyopic	hypermetrope	yes	normal	no-lenses
presbyopic	myope	no	reduced	no-lenses
presbyopic	myope	no	normal	no-lenses
presbyopic	myope	yes	reduced	no-lenses
presbyopic	myope	yes	normal	hard-lenses
presbyopic	hypermetrope	no	reduced	no-lenses
presbyopic	hypermetrope	no	normal	soft-lenses
presbyopic	hypermetrope	yes	reduced	no-lenses
presbyopic	hypermetrope	yes	normal	no-lenses

The (final) neurules produced are shown in Table 6. For the first two initial neurules, the calculated factors successfully classified all training patterns. The produced neurules NR1 and NR2. However, the same didn't happen with the third initial neurule. Its training set had to be split, two copies of the third initial neurule were trained with each subset and neurules and finally neurules NR3 and NR4 were produced.

Table 6 The neurules produced from the empirical set of lenses problem

NR1	NR3
(-13.1) **if** age is young (8.8),	(-4.6) **if** age is young (-4.4),
age is pre-presbyopic (1.5),	age is pre-presbyopic (2.6),
age is presbyopic (1.2),	age is presbyopic (3.2),
spectacle is myope (1.6),	spectacle is myope (-4.2),
spectacle is hypermetrope (-2.7),	spectacle is hypermetrope (3.4),
astigmatic is no (-6.1),	astigmatic is no (-4.5),
astigmatic is yes (4.4),	astigmatic is yes (3.3),
tear-rate is reduced (-5.7),	tear-rate is reduced (6.5),
tear-rate is normal (4.6)	tear-rate is normal (-8.0)
then lenses-class is hard-lenses	**then** lenses-class is no-lenses
NR2	**NR4**
(-14.6) if age is young (6.4),	(-2.2) if age is young (-2.6),
age is pre-presbyopic (6.9),	age is pre-presbyopic (-2.5),
age is presbyopic (-0.4),	age is presbyopic (5.0),
spectacle is myope (-3.9),	spectacle is myope (1.0),
spectacle is hypermetrope (3.1),	spectacle is hypermetrope (-2.5),
astigmatic is no (6.9),	astigmatic is no (5.1),
astigmatic is yes (-7.4),	astigmatic is yes (-6.2),
tear-rate is reduced (-7.9),	tear-rate is reduced (8.1),
tear-rate is normal (6.2)	tear-rate is normal (-9.5)
then lenses-class is soft-lenses	**then** lenses-class is no-lenses

6 Efficient Updating of a Neurule Base Produced from Empirical Data

In certain domains, training examples become available over time. Therefore, an aspect of interest involves the efficient updates of a NRB to reflect availability of new empirical source knowledge. In (Prentzas and Hatzilygeroudis 2007b) we present an efficient mechanism for performing such updates. The mechanism is based on splitting trees containing information regarding the splitting process for each training set of each initial neurule, in a similar way to that in Section 4. The root of each tree corresponds to the initial training set. Descendant nodes correspond to the subsequent subsets into which the initial training set was split. Each leaf denotes subsets that was successfully trained and produced a neurule. The members of the LCP that guided each split are attached to the corresponding branches of the tree.

The splitting tree is useful to perform updates in case more than one sibling neurules have been produced. The availability of a new training example means insertion of a new success example into a specific initial training set and insertion of a new failure example into all other initial training sets. Splitting trees enable to perform such updates efficiently.

To insert a new success example not satisfied by the existing neurules produced from the initial training set, the corresponding splitting tree is traversed to starting from the root and ending at a leaf or an intermediate node. Traversing is based on the closeness between the new success example and the LCPs attached to the edges of the splitting tree. Retraining of the corresponding subset is performed updating the NRB and the splitting tree.

The insertion of a failure example not satisfied by the existing neurules into an initial training set requires training of the subsets corresponding to leaves of the splitting tree whose corresponding neurules misclassify the new failure example. The corresponding existing neurules are removed from the NRB, whereas the newly created ones are inserted.

7 Neurule-Based Inference Engine

The *neurule-based inference engine* implements the way neurules co-operate to reach a conclusion. The choice of the next rule to be considered is based on a neurocomputing measure, but the rest is symbolic (Hatzilygeroudis and Prentzas 2010).

Generally, the output of a neurule is computed according to Eq. (1) (Section 2.1). However, it is possible to deduce the output of a neurule without knowing the values of all of its conditions. To achieve this, we define for each neurule the *known sum (kn-sum)* and the *remaining sum (rem-sum)*. More specifically, 'known-sum' is the weighted sum of the values of the already known (i.e. evaluated) conditions (inputs) of the corresponding neurule. 'Remaining sum' is the sum of absolute values of significance factors corresponding to all unevaluated conditions of the neurule. Therefore, 'remaining sum' represents the largest possible weighted sum of the remaining (i.e. unevaluated) conditions of the neurule.

If |*kn-sum*| > *rem-sum* for a certain neurule, then evaluation of its conditions can stop, because its output can be deduced regardless of the values of the unevaluated conditions. In this case, its output is guaranteed to be '1' if *kn-sum* > 0 whereas it is '-1', if *kn-sum* < 0. In the first case, we say that the neurule is *fired*, whereas in the second that it is *blocked*.

A condition evaluates to 'true', if it matches a fact in the working memory that is, there is a fact with the same variable, predicate and value. A condition evaluates to 'unknown', if there is a fact with the same variable, predicate and 'unknown' as its value. A condition cannot be evaluated if there is no fact in the working memory with the same variable. In this case, either a question is made to the user to provide data for the variable, in case of an input variable, or an intermediate neurule with a conclusion containing the variable is examined, in case of an intermediate variable. A condition with an input variable evaluates to 'false', if there is a fact in the working memory with the same variable, predicate and different value. A condition with an intermediate variable evaluates to 'false' if additionally to the latter there is no unevaluated intermediate neurule that has a conclusion with the same variable. Inference stops either when one or more output neurules are fired (success) or there is no further action (failure). To facilitate inference, conditions of neurules are organized according to the descending order of their significance factors.

In (Hatzilygeroudis and Prentzas 2001a) we present initial work for the provision of explanations concerning neurule-based inference. Explanations involve 'how' type rules justifying how conclusions were reached.

Neurule-based inference has certain advantages. When a neurule base is produced from a symbolic rule base, experimental results have shown that neurule-based inference is more efficient than the corresponding symbolic rule-based inference (Hatzilygeroudis and Prentzas 2000b). Another advantage of neurule-based reasoning compared to symbolic rule-based reasoning is the ability to reach conclusions from neurules even if some of the conditions are unknown. This is not possible in symbolic rule-based reasoning. A symbolic rule needs all its conditions to be known in order to produce a conclusion.

Most neuro-symbolic approaches, except connectionist expert systems, do not support functionalities like interactive inference and provision of natural explanations. Neurule-based inference is more efficient than the inference mechanism used in connectionist expert systems (Hatzilygeroudis and Prentzas 2010).

8 Combining Neurule-Based and Case-Based Reasoning

Case-based reasoning is an approach that exploits knowledge encompassed in stored past cases to handle similar new cases (Aamodt and Plaza 1994). It is useful in several domains where an abundant number of past cases are available. A case-based system stores useful experience obtained when handling new cases and is continuously enhanced during operation. Case-based representations offer several advantages such as easy knowledge acquisition, naturalness, modularity, ability to express specialized knowledge, self-updatability. There are also issues of CBR that may give rise to problems such as adaptation, inference efficiency regarding case retrieval, provision of explanations, difficulties in knowledge acquisition in certain domains (Prentzas and Hatzilygeroudis 2007a, 2009).

Combinations of case-based reasoning with other intelligent methods have been pursued in several domains resulting into more effective representation schemes. One of the most effective types of combinations involves combination of case-based reasoning with rule-based reasoning (Prentzas and Hatzilygeroudis 2007a). Such combinations offer benefits since the advantages of rule-based reasoning and case-based reasoning are complementary to a large degree. An overall advantage of such combined approaches involves naturalness and modularity of the representation scheme. Neurules are a type of integrated rules offering advantages compared to symbolic rules. In (Prentzas and Hatzilygeroudis 2002, Hatzilygeroudis and Prentzas 2004c) we explored the combination of neurule-based with case-based reasoning. A main benefit, among others, deriving from this combination concerns accuracy improvement as cases may fill in gaps of neurules in domain knowledge representation. Furthermore, characteristics of the formalism involve naturalness, modularity, ability to exploit multiple types of knowledge sources and self-updatability. Few approaches combine case-based reasoning with multiple other intelligent methods with the other methods being outside the case-based reasoning cycle.

In the representation scheme combining neurules with case-based reasoning, neurules index cases representing their exceptions. The indexing construction module implements the process of acquiring an indexing scheme. The specific process may take as input alternative types of available knowledge: (a) available neurules and cases or (b) available symbolic rules and exception cases.

Let us consider that the indexing process takes as input available neurules and cases. To acquire an indexing scheme, neurule-based reasoning is performed for the neurules based on the attribute values of each case. A case is indexed as a neurule's exception, whenever the neurule fires and the value of the conclusion variable do not match the corresponding attribute value of the case.

The alternative type of knowledge concerns an available formalism of symbolic rules and indexed exception cases as the one presented in (Golding and Rosenbloom 1996). The indexing scheme is acquired by first converting symbolic rules to neurules and then associating the produced neurules with the exception cases of the symbolic rules belonging to their merger sets.

The hybrid inference process combining neurule-based with case-based reasoning focuses on neurules (i.e. neurule-based reasoning). If an adequate number of the conditions of a neurule are fulfilled so that it can fire, firing of the neurule is suspended and CBR is performed for its indexed exception cases. CBR results are evaluated as in (Golding and Rosenbloom 1996) to assess whether the neurule will fire or whether the conclusion proposed by the exception case will be considered valid.

9 Conclusions

In this paper, we present an overview of our main research work involving neurules, a type of hybrid neuro-symbolic rules. An attractive feature of neurules is that compared to other connectionist approaches they retain the modularity and to some degree the naturalness of symbolic rules. In contrast to most neuro-symbolic approaches, a neurule-based system also provides an interactive inference mechanism and explanation facilities. We outlined aspects regarding construction of neurules from symbolic rule bases or empirical data, efficient updating of a neurule base constructed from symbolic rule bases or empirical data, neurule-based inference and combination of neurules with case-based reasoning.

Neurules have been used in developing an Intelligent Tutoring System (Prentzas, Hatzilygeroudis and Garofalakis 2002, Hatzilygeroudis and Prentzas 2004b). Intelligent Tutoring Systems (ITSs) require discrete knowledge bases to perform tasks of their different units (i.e. user modeling unit, pedagogical unit). Neurules facilitated the development and performance of the ITS since they satisfy most of the representation requirements concerning ITSs (Hatzilygeroudis and Prentzas 2004d, 2006). More specifically, neurule bases can be constructed from alternative knowledge sources producing a natural and modular representation scheme. Incremental development of neurule bases is also supported to accommodate source knowledge changes. Furthermore, neurule-based inference is natural, robust and time-efficient.

Our future work is directed to a number of aspects. Such aspects involve finding ways to (a) improve the neurule-based inference efficiency, (b) provide natural explanations, (c) incorporate fuzziness into neurules and (d) improve the mechanisms constructing neurules. Another future direction will involve use of neurules in different applications.

References

Aamodt, A., Plaza, E.: Case-based reasoning: foundational issues, methodological variations and system approaches. AI Communications 7, 39–59 (1994)

Andrews, R., Diederich, J., Tickle, A.: A survey and critique of techniques for extracting rules from trained artificial neural networks. Knowledge-Based Systems 8, 373–389 (1995)

Bader, S., Hitzler, P.: Dimensions of neural-symbolic integration – a structured survey. In: Artemov, S., Barringer, H., d'Avila Garcez, A.S., Lamb, L.C., Woods, J. (eds.) We Will Show Them: Essays in Honour of Dov Gabbay. International Federation for Computational Logic, vol. 1, pp. 167–194. College Publications (2005)

Bader, S., Hitzler, P.: Holldobler Connectionist model generation: a first-order approach. Neurocomputing 71, 2420–2432 (2008)

Bookman, L., Sun, R. (eds.): Special issue on integrating neural and symbolic processes. Connection Science, vol. 5(3-4) (1993)

Browne, A., Sun, R.: Connectionist inference models. Neural Networks 14, 1331–1355 (2001)

Cloete, I., Zurada, J.M. (eds.): Knowledge-based neurocomputing. The MIT Press, Cambridge (2000)

Frank, A., Asuncion, A.: UCI Machine Learning Repository, School of Information and Computer Science, University of California, Irvina, CA (2010), http://archive.ics.uci.edu/ml(accessed October 9, 2010)

Fu, L.M.: Knowledge-based connectionism for revising domain theories. IEEE Transactions on Systems, Man, and Cybernetics 23, 173–182 (1993)

Fu, L.M. (ed.): Proceedings of the International Symposium on Integrating Knowledge and Neural Heuristics. Pensacola, Florida (1994)

Gallant, S.I.: Connectionist expert systems. Communications of the ACM 31, 152–169 (1988)

Gallant, S.I.: Neural network learning and expert systems. The MIT Press, Cambridge (1993)

d'Avila Garcez, A.S., Broda, K., Gabbay, D.M.: Neural-symbolic learning systems: foundations and applications. In: Perspectives in Neural Computing. Springer, Heidelberg (2002)

d'Avila Garcez, A., Gabbay, D., Holldobler, S., Taylor, J.: Special issue on neural-symbolic systems. Journal of Applied Logic 2 (2004)

d'Avila Garcez, A., Lamb, L.C., Gabbay, D.M.: Connectionist modal logic: representing modalities in neural networks. Theoretical Computer Science 371, 34–53 (2007)

Ghalwash, A.Z.: A recency inference engine for connectionist knowledge bases. Applied Intelligence 9, 201–215 (1998)

Golding, A.R., Rosenbloom, P.S.: Improving accuracy by combining rule-based and case-based reasoning. Artificial Intelligence 87, 215–254 (1996)

Gonzalez, A., Dankel, D.: The engineering of knowledge-based systems: theory and practice. Prentice-Hall, Upper Saddle River (1993)

Hatzilygeroudis, I., Prentzas, J.: Neurules: integrating symbolic rules and neurocomputting. In: Fotiades, D., Nikolopoulos, S. (eds.) Advances in Informatics. World Scientific Publishing, Singapore (2000a)

Hatzilygeroudis, I., Prentzas, J.: Neurules: improving the performance of symbolic rules. International Journal on AI Tools 9, 113–130 (2000b)

Hatzilygeroudis, I., Prentzas, J.: An efficient hybrid rule based inference engine with explanation capability. In: Kolen, J., Russell, I. (eds.) Proceedings of the Fourteenth International Florida Artificial Intelligence Research Society Conference. AAAI Press, Menlo Park (2001a)

Hatzilygeroudis, I., Prentzas, J.: Constructing modular hybrid knowledge bases for expert systems. International Journal on AI Tools 10, 87–105 (2001b)

Hatzilygeroudis, I., Prentzas, J.: Neuro-symbolic approaches for knowledge representation in expert systems. International Journal on Hybrid Intelligent Systems 1, 111–126 (2004a)

Hatzilygeroudis, I., Prentzas, J.: Using a hybrid rule-based approach in developing an intelligent tutoring system with knowledge acquisition and update capabilities. Expert Systems with Applications 26, 477–492 (2004b)

Hatzilygeroudis, I., Prentzas, J.: Integrating (rules, neural networks) and cases for knowledge representation and reasoning in expert systems. Expert Systems with Applications 27, 63–75 (2004c)

Hatzilygeroudis, I., Prentzas, J.: Knowledge representation requirements for intelligent tutoring systems. In: Lester, J.C., Vicari, R.M., Paraguaçu, F. (eds.) ITS 2004. LNCS, vol. 3220, pp. 87–97. Springer, Heidelberg (2004d)

Hatzilygeroudis, I., Prentzas, J.: Knowledge representation in intelligent educational systems. In: Ma, Z. (ed.) Web-based intelligent e-learning systems: technologies and applications. Idea Group Inc., Hershey (2006)

Hatzilygeroudis, I., Prentzas, J.: Integrated rule-based learning and inference. IEEE Transactions on Knowledge and Data Engineering 22, 1549–1562 (2010)

Haykin, S.: Neural Networks and Learning Machines. Prentice Hall, Upper Saddle River (2008)

Hilario, M.: An overview of strategies for neurosymbolic integration. In: Sun, R., Alexandre, E. (eds.) Connectionist-symbolic integration: from unified to hybrid approaches. Lawrence Erlbaum Associates, Mahwah (1997)

Holldobler, S., Kalinke, Y.: Towards a massively parallel computational model for logic programming. In: Proceedings of ECAI 1994 Workshop on Combining Symbolic and Connectionist Processing. pp. 68–77. ECCAI (1994)

Komendantskaya, E., Lane, M., Seda, A.K.: Connectionist representation of multi-valued logic programs. In: Hammer, B., Hitzler, P. (eds.) Perspectives of Neural-Symbolic Integration. Springer, Heidelberg (2007)

McGarry, K., Wermter, S., MacIntyre, J.: Hybrid neural systems: from simple coupling to fully integrated neural networks. Neural Computing Surveys 2, 62–93 (1999)

Medsker, L.R.: Hybrid intelligent systems. Kluwer Academic Publishers, Dordrecht (1995)

Prentzas, J., Hatzilygeroudis, I.: Integrating hybrid rule-based with case-based reasoning. In: Craw, S., Preece, A.D. (eds.) ECCBR 2002. LNCS (LNAI), vol. 2416, p. 336. Springer, Heidelberg (2002)

Prentzas, J., Hatzilygeroudis, I., Garofalakis, J.: A web-based intelligent tutoring system using hybrid rules as its representational basis. In: Cerri, S.A., Gouardéres, G., Paraguaçu, F. (eds.) ITS 2002. LNCS, vol. 2363, p. 119. Springer, Heidelberg (2002)
Prentzas, J., Hatzilygeroudis, I.: Rule-based update methods for a hybrid rule base. Data and Knowledge Engineering 55, 103–128 (2005)
Prentzas, J., Hatzilygeroudis, I.: Construction of neurules from training examples: a thorough investigation. In: Garcez, A., Hitzler, P., Tamburini, G. (eds.) Proceedings of the ECAI 2006 Workshop on Neural-Symbolic Learning and Reasoning (2006)
Prentzas, J., Hatzilygeroudis, I.: Categorizing approaches combining rule-based and case-based reasoning. Expert Systems 24, 97–122 (2007a)
Prentzas, J., Hatzilygeroudis, I.: Incrementally updating a hybrid rule base based on empirical data. Expert Systems 24, 212–231 (2007b)
Prentzas, J., Hatzilygeroudis, I.: Combinations of case-based reasoning with other intelligent methods. International Journal of Hybrid Intelligent Systems 6, 189–209 (2009)
Reichgelt, H.: Knowledge representation, an AI perspective. Ablex, New York (1991)
Souici-Meslati, L., Sellami, M.: Toward a generalization of neuro-symbolic recognition: an application to Arabic words. International Journal of Knowledge-based and Intelligent Engineering Systems 10, 347–361 (2006)
Sun, R., Alexandre, E. (eds.): Connectionist-symbolic integration: from unified to hybrid approaches. Lawrence Erlbaum Associates, Mahwah (1997)
Teng, T.-H., Tan, Z.-M., Tan, A.-H.: Self-organizing neural models integrating rules and reinforcement learning. In: Proceedings of the IEEE International Joint Conference on Neural Networks, IEEE, Los Alamitos (2008)
Towell, G., Shavlik, J.: Knowledge-based artificial neural networks. Artificial Intelligence 70, 119–165 (1994)
Wermter, S., Sun, R. (eds.): Hybrid neural systems. Springer, Heidelberg (2000)
Xianyu, J.C., Juan, Z.C., Gao, L.J.: Knowledge-based neural networks and its application in discrete choice analysis. In: Proceedings of the Fourth International Conference on Networked Computing and Advanced Information Management. IEEE Computer Society Press, Los Alamitos (2008)
Yu, L., Wang, L., Yu, J.: Identification of product definition patterns in mass customization using a learning-based hybrid approach. International Journal of Advanced Manufacturing Technologies 38, 1061–1074 (2008)

Author Index